# THE SECRET LIFE OF CONSERVATIONISTS

# The Secret Life of Conservationists

*A compilation of stories by Lonely Conservationists*

## COMPILED BY JESSIE PANAZZOLO

*Edited by Renuka Kulkarni*

Lonely Conservationists

# ARTISTS

*Cover art by Jack O'Connor*

*Illustrations by Phalguni Ranjan*

To any lonely conservationist, may you now feel less alone.

# CONTENTS

# | 1 |

# Foreword

## JESSIE PANAZZOLO

Conservation has traditionally been glorified in mainstream and social media as a job that involves beautiful and exotic animals in breathtaking corners of the globe. Thanks to the likes of David Attenborough's awe-inspiring documentaries and an industry of people posting snapshots of their experiences in far off places online, the public perception of working to conserve native species is all joy, beauty and wonder.

Three years ago, I truly believed that I was the only one who had missed out on experiencing the financial and professional successes of being a conservationist, but it turns out that I was extremely wrong about that. With nothing to lose, I shared my frustrations online in a blog that I titled *Lonely Conservationists*, urging anyone who felt the same to share their stories too. There are so many lonely conservationists in fact, that from the day I posted my very first blog to www.lonely-conservationists.com, until the time of writing this book (2021), I have been able to share a diverse and worldwide experience of navigating the industry every single week. *Lonely Conservationists* is now a thriving online community for conservationists to share their stories about the

1

not-so-glamorous aspects of the industry and has become a sanctuary of like-minded friends, collaborators and supporters.

This book aims to share the global perspectives of nature conservationists through their personal and authentic stories. These stories have been adapted from those that were once submitted to the blog, and as a result, span over three years. They cover the challenges of being a conservationist in 2019- a time before the world seemed truly dystopian. It explores the Australian bushfires, the global COVID-19 pandemic and the Black Lives Matter movement of 2020, as well as the emergence of a new normal in 2021. These stories aren't placed in any particular timeline and some have been updated to show how the author's lives have changed since they wrote their initial story.

Understandably, it may be challenging to digest so many stories of hardship at once, so feel free to read this book as you can, and without the pressure to rush through. I would also like to issue a warning upfront that some of these stories contain themes of mental illness, death and sexual assault.

Last but definitely not least, I'd like to acknowledge the first conservationists of the place that you are reading from. For me, I am writing on Wurundjeri country within the Kulin Nation. I would like to extend my acknowledgement into thanks for all that was done and all that continues to be done by First Nations Peoples to conserve natural spaces and ecosystems across the globe.

We have a lot to learn by listening to those around us, and I'm hoping that by sharing these stories from a range of diverse voices, we can remember the values of inclusion, collaboration and community and their importance in succeeding in our conservation efforts.

Without further ado, here are the stories that revolutionised the narrative of what it means to be a conservationist.

# | 2 |

# Finding your identity in a crowded sector

## JAMIE SNEDDON

My name is Jamie. I'm a zoologist/conservationist/field biologist/ ecologist/scientist/clueless graduate with – surprise, surprise – an identity crisis. The issue of who/what I am is something that I've been struggling with for much of my life and throughout my entire career.

While most people don't like being pigeonholed, there's a sense of security to be found in knowing your place in a chaotic world. Being able to define yourself with a neat, irrefutable job title is something that I think people in some industries take for granted. One of the first things people ask you is 'What do you do for a living?'.

For many, the answer is simple. *I'm a teacher. I'm a personal trainer. I'm a lion tamer.*

For me, my answer changes depending on my audience or even my mood. For most of us, this kind of identity crisis didn't hit until we entered the world of work. In my case, it began much before that.

I've always felt like I needed to try harder than others when it came to education. At primary school, it was a constant battle to get me to focus. Windows were my downfall; why should I listen to the teacher when I could watch the crows outside? If I was sat away from the windows, I would just collect insects in my pencil case at break and take them into class with me. Problem solved.

My lack of focus was amplified when it came to maths. If I was in school right now, I'm almost certain that I would be diagnosed with some kind of dyscalculia. To put it simply, numbers just didn't make sense to me. I distinctly remember thinking of numbers as male and female: 1,3,6,9,11, etc. were female and 2,4,5,7,8, etc. were male. It was odd and didn't help me with my maths at all, but I think it was perhaps my mind trying to rationalise something I couldn't understand.

As time went on, I fell behind more and more. When I couldn't recite my times-tables, I would be held in at break and lunch and made to recite them in front of all the teachers in the staff room. In hindsight, that's a pretty cruel thing to do to a struggling child. I would always manage it, even if it was very strained, and immediately forget it all again once I left the room. This didn't help me with maths; it did quite the opposite. It gave me painful insecurities and anxiety when it comes to anything number related. This still affects me today and is a real hurdle in my battle to be taken seriously in the conservation sector.

I made it through secondary school with the same issues but managed to get the grades I needed to study zoology at university. Bring on the practical education and the freedom to learn about topics that interest me. Wrong! Welcome to the world of academia.

The problems started early on. I dropped higher chemistry in secondary school because I found it too challenging. You can imagine my horror when advanced chemistry was a mandatory course in my first year of university. I thought it'd be okay because I could finally get stuck into fieldwork, right? Wrong again! Hope you enjoy windowless lecture theatres and labs under harsh artificial lights!

Maybe I was naïve; correction, I was *definitely* naïve to think that you had to spend time in the natural world to learn about the natural world.

Coming to a city after living in the Scottish Highlands, I felt trapped. Concrete everywhere, no forests, no escaping to the mountains. I volunteered with red squirrel monitoring projects in parks and captured footage of urban foxes on a cheap camera trap. It was these optional outdoor activities that kept me sane. I would have expected others in my course to feel the same way, but many complained about the rare few fieldwork sessions that we did have. Searching for rabbit poo in the snow was apparently a pointless exercise for some zoology students!

Snow is great and poo is great. Period.

For the most part, I managed to do quite well at university. Although I still felt that I had to work far harder than my classmates, I didn't struggle too much overall. That is until statistics came along.

If it wasn't for my friends, I would have never passed this element of my degree. It made me feel stupid and a complete fraud. If I couldn't do statistics or even simple mathematics, how could I be a conservationist? This feeling was amplified when a lab assistant shamed me in front of my lab partners. I didn't understand a part of an equation and she proceeded to patronise me by saying, very slowly so my "simple" brain could keep up:

*"If you have x number of apples and I take away x number of apples, how many apples are you left with?"*

Humiliating. As a third-year science student, I didn't think people would make me feel like dirt and certainly wouldn't publicly humiliate me. That took a while to bounce back from.

Amazingly, despite my fight or flight reaction to basic maths, I graduated with a 2:1 B.Sc. in Zoology. Time to face the world of work!

After attending a career day in my fourth year, I think most of my year felt dull. It was full of people who had volunteered for 5 years and had just secured a placement or had managed national parks in the Middle East but moved to the UK and had to volunteer before securing paid work. Very inspiring stuff! While everyone else felt demoralised, it made me determined to break the mould. Surely these people just weren't trying hard enough or thinking outside the box!

It was this logic that led me to become a falconer.

*"Guys, do you know there are companies that fly birds of prey as a form of non-lethal pest control? Mental right?!"*

It wasn't conventional conservation, but it involved monitoring wild birds and working with animals. Most importantly, it was paid work. The problem was that after applying three times, I still hadn't secured an interview. My obvious lack of experience flying birds of prey was going to be an issue. To overcome the usual application chain, which clearly wasn't working, I found an employee, got them to give me the number of the boss and got in touch with him.

Straight to source thinking. I met the man while he was flying a peregrine falcon from the top of a multi-storey car park, and convinced him to give me a chance. Whatever I said worked, because the next thing I knew, I was the proud owner of Maggie, a harris hawk, and running around Scotland, keeping schools, nuclear power plants and construction sites clear of seagulls.

Now a word of warning: this was not a dream job.

Birds of prey are hard to work with. They don't like people all that much and can be very temperamental. Having to pull Maggie's talons out of my girlfriend's face taught me that.

Despite some rocky moments, I found the birds great company. I was alone 95% of the time, isolated by the strange nature of my job and feeling a bit lost. I enjoyed watching gull colonies and learning about seabird ecology, but I also had to do my fair share of unpleasant work. For example, I had to fly Maggie on highstreets to keep seagulls away at lunchtime, a completely pointless job because seagulls are too intelligent for this kind of non-lethal pest control. The result: Maggie being mobbed by gulls and me being spat at by drug addicts because I wouldn't let them hold her. Fun time.

Having to check rat boxes at waste treatment centres was another less-than-glamorous element of my job. But it wasn't being spat on or having rats run up my arm that broke me; it was the day I was forced to remove three seagull chicks from a roof.

Now when you have birds of prey, you have to kill animals at some point. It's not something I took pleasure in; unlike most people, I admire seagulls. But if your hawk catches a gull and can't finish the job, then you need to humanely dispatch it before your hawk ends up with a broken wing. Gulls fight back and are built like tanks. I could deal with this unsavoury element of the job because it happened rarely.

But on that day, things came to a head. A little clutch of gull chicks, maybe a week from fledging, were sitting on the remaining sections of roof left during a demolition project. I had argued for weeks that the chicks were about to fledge so we should leave them alone. Unfortunately, the business pushed for removal and I was told to remove the chicks or lose my job.

I removed the chicks.

Each one had a different personality: one ran, one fought back, and one just sat there and accepted its fate.

I went home, grey-faced, sat in the shower on the verge of tears and quit my job a few weeks later. It was needless killing. I was young, naïve and scared to lose my income. It's something that I feel a deep shame about to this day. If I could go back, I would have left the chicks alone and quit on the spot. It's easier to have morals when you don't have to fear being homeless if you lose your job. From a zoology graduate to this. Not exactly the David Attenborough-esque future I had imagined.

Following this job and a slight detour into retail - thanks to my swift exit from flying birds - I decided that I needed to get back to actual conservation.

This meant volunteering. I've volunteered with squirrels, wildcats and even some birds and fish, short term. The overall theme seems to be learning to do very skilled work, doing that work for a long time and getting no money for your trouble. Volunteering short term seems to spiral into years and it's not long before your sense of self-worth is whittled away. What you're left with is a very skilled and hardworking individual, but also someone without an ounce of self-respect.

This was perfectly summed up when I was hit with norovirus during a particularly intense volunteering post. I'm not exaggerating when I say that I thought the virus was going to kill me. Despite this, I only took a few days off because I had so much *work* to do. What happens when you stumble around the woods for hours on end when you haven't eaten in days and are still reeling from an aggressive virus? You collapse multiple times, wake up very confused, stumble to your feet and keep going.

People who haven't tried to break into the conservation sector don't understand this kind of behaviour. I don't think it's normal, but this sector fosters this insane mindset. Nothing you ever do is good enough, you can always work harder and there are a million other people who would volunteer in your post if you can't cope. Those desperate, hungry graduates nipping at your heels and the usual promise of an elusive job appearing at the end of your struggle are enough to keep you going for a while. The promised jobs don't appear that often.

Sometimes you get a compliment and a nice mug, but not paid employment. I've literally managed a conservation project for free, had great success doing so and still the job I was promised is yet to materialise. The problem is that while you're in the volunteer maze, you hold on to the logic that if you leave, it will all have been for nothing. That if you can just hold on a bit longer, you'll be rewarded with the breakthrough you've been working towards. You should be grateful just to have gotten this far. It's a sad mindset and eventually becomes pretty pathetic. How can you expect someone to respect you when you don't respect yourself?

The day I refused to volunteer anymore was one of the best things I think I could have done, for my conservation career and my mental health.

Kim, my girlfriend, couldn't take seeing me work myself to death for no reward anymore. I had no money and wanted to move on with my life. I wanted to get a dog, stop renting flats with horrible landlords and think about having a family. Conservation couldn't offer me these things and Kim, with her stable teacher wage, deserved to live the life we both wanted. She had worked hard, made good choices and had supported me every step of the way. Eventually, I couldn't cope with the feeling that my hopeless career was holding her back. So, I went to work in a school as a pupil support assistant, the most stable job I could find at the time. I made some money, started to move on with my life and I left conservation behind to save what was left of my passion. If conservation wouldn't pay me for my hard work, then I wasn't going to hold on forever.

But it wasn't long before I started to yearn for the forests again.

Before working in the school, I could walk around and find squirrels by the sound of pinecones being nibbled in the canopy. The forest was a tranquil place that heightened all my senses. Now they were overwhelmed.

Bells were deafening, the lights were bright, the air heavy and stagnant. It was an alien world.

Despite feeling like a fish out of water, I found out that I was actually very good at my job. Being a young man, in a female-dominated workforce, meant that I helped out a lot of angry boys who needed a good male role model. I taught a lot of kids about nature and even got them to take part in beach cleans.

On good days, it was great. But there were days when I just wanted to run away. If you've ever had a child call you a 'f-king moron!', you'll know what I mean. The thought that I had spent four years in university for this crossed my mind a lot. Essentially, I was back to feeling underappreciated and alone, but in a completely different way. Colleagues spoke about the latest soap drama or the fact that their neighbour never cuts the grass. Riveting stuff. Everyone had an air of being unhappy and bored, but no one seemed to be doing anything to escape this personality vacuum.

In a world where you're defined by your career, I felt judged constantly. People would ask why I worked in a school if I had a zoology degree. I thought the same thing, so it was hard to justify. When I volunteered, people were confused by my choices, but hey, at least I was interesting. I enjoyed having stories to tell: the time a reindeer had kicked me and left me hobbling for weeks, or the time a half-awake and very angry cat had escaped in my car.

People would give me pitying looks when I told them that I worked in a school. I would avoid the topic of work at all costs and felt sorry for Kim. She could be judged just as easily for being attached to a guy who was either a hippie, running around the woods for free, or someone who couldn't cut it in his sector and hence took a low paid position in a school. I would apply for jobs constantly and get knocked back, time and time again.

Each time, the look of pity from people grew stronger. There I was: a failure. People didn't understand the sector, so it wasn't the industries fault. It was mine. Years had gone by, people had moved up and progressed, but there's Jamie still struggling and getting nowhere.

Then, at long last, out of nowhere, one of the promised jobs appeared!

When I left my volunteer post with the squirrel project in 2017, I told myself that I wouldn't volunteer for anything again. I was worth more than that and until I put my foot down, I wasn't going to get anywhere. Two years later, at my lowest point, I was offered a paid job on that same squirrel project and that's where I find myself now.

I am Jamie, the professional squirrel tracker.

Although I'm still not sure if that's exactly what I can call myself.

My job title is research field assistant, but I still go by ecologist/field biologist/zoologist, etc. depending on the audience. I still haven't found my identity but it's a good start. Working with volunteers has also allowed me to appreciate how far I've come. One volunteer told me that people in her position look at me as a success story, a strange concept for my warped sense of self.

The work is hard, the pay isn't great, and my work-life balance is still poor but at least Kim doesn't need to watch out for hawk attacks anymore. Being on a zero-hour contract means that the job hunt continues but for now, I'm a paid member of the highly competitive conservation sector and that feels pretty good.

The advice I would give to anyone is to know your worth. Many of your bosses will diminish your skills because they want you to keep working for free. People with self-worth move on, they grow, they achieve. Volunteering is an essential element of the industry we work in but you should know when to stop. Learn what you need to and move on. Don't hold out hope for a job that might never appear. Network, speak to people honestly and be you. Your personality can be just as valuable as your skills but what makes you unique won't necessarily shine through on your CV, no matter how many times you rewrite it.

Overall, just remember that your love for conservation has to balance with your love for yourself and the people in your life. If you're a penniless, sleep-deprived, broken soul then you're no use to anyone. Take comfort in the fact that you're not alone and there are thousands of crazy conservationists out there that understand you.

We're all in this together.

## Update: summer 2021

My name is Jamie, and I've found my place in the crowded conservation sector. In the coming years, I'll be working with a team of amazing professionals to save the Scottish wildcat, a once-in-a-lifetime opportunity.

The above story isn't the most positive read as I was in a fairly low and frustrating place when it was first written, but I hope it can help others to see the light at the end of the tunnel. Succeeding in this industry means networking, hard work and a large dose of being in the right place at the right time. Take opportunities when they arise, but know when to move on to new pastures. One day you'll wake up and realise that you don't need to fight, scrape and beg for a job anymore. Our broken world needs conservationists now more than ever so don't give up! Your dream job could be right around the corner.

# | 3 |

# My current mental state

## MACIE EDWARDS

A friend recently sent me an article about an online community called *Lonely Conservationists* (LC), a group that I was surprised to have never heard of before, as the name itself was immediately relatable. I read that article with a tight throat, choking back tears. It felt so reflective of what I had been dealing with in this field and the challenges that I am still facing. While it is relieving and encouraging to know that I am not alone in this deepening state of loneliness and frustration, it does not change reality.

I decided on this career path when I was in high school. I had a vague idea of what I wanted to do. I loved animals, the outdoors, and Steve Irwin. I wanted the privilege and honour of being a National Park Ranger. I wanted to make a difference and do something that felt worthwhile. I wanted to contribute to the protection and better under-standing of our natural world. I entered this career with a passion that has since been slipping away.

I interned for the National Park Service for one season (May-Octo-ber), and in that short time, I had my rose-tinted view of the institu-tion completely wiped away. I made roughly $3.25 (USD) an hour and needed to work a second job for a couple of months to earn some ex-tra money. I wasn't the only one who did this. Others in the program collected food stamps to help pay for groceries. At this time, the park didn't even bother to give us the dignity of a uniform that would have made us recognisable as park employees to the hundreds of visitors we came in contact with daily. *"Do you even work here?"* was a question we frequently answered.

I also came to realise how political the institution is, how difficult and confusing the hiring process can be, and how competitive it is. It can take years of seasonal work to even qualify for a permanent posi-tion. I have met many underpaid, undervalued, and jaded employees. *"We get paid in sunsets"* is a phrase I've heard or read many times now, more often than not with a sharp tone of cynicism behind it.

I have since given up on my dream of wearing the NPS uniform. It feels unattainable and no longer worth it.

The majority of field jobs are seasonal, each typically lasting for six months. This vagabond, rootless, constantly fluctuating lifestyle is one I wish that I was better at. But if I'm being honest with myself, it is a way of living that is slowly eating away at my well-being. I live in six-month increments. I work in the summer, then live with my parents in the winter, taking whatever part-time, table-waiting job I can find until another summer season starts. All of my friends are long-distance and scattered around the country because most are in the same line of work that I am. I know that it is very unlikely that I will ever see many of them again as life continues to pull us in different directions.

I feel like a failure because I am unemployed (again). I am uncertain of my future. And I am so sick and tired of having to answer the same three questions over and over again. *"How's the job search going?"*, *"Find a job yet?"*, *"Why don't you just get a permanent job?"*. People even wonder why I don't just get a seasonal job in the winter somewhere down south. I have to explain that it's not that easy, and that, believe it or not, despite the stress and anxiety, I feel blessed to have that kind of quality time with my family while I can.

Not many people know about my first ever panic attack. It happened last winter while driving home from my brother's house, one week before I left home again for another season in North Dakota which is 2,000 miles (approx. 3,200 km) away. I was already crying, upset about my career, leaving my family again and everything I miss out on when I'm gone. The uncertainty of where it is all going, the chronic stress from months of unemployment, feeling like a failure for living with my parents to save money while I look for the next job

...and then I started hyperventilating.

My entire body began to tingle and go numb. There was a black vignette forming around my vision. Now, mind you, this was all happening while I was speeding down the interstate, still thirteen miles (approx. 20 km) away from my exit and trapped in the left-hand lane next to a semi (a kind of trailer truck). It was not ideal, to say the least.

I come from a family of police officers and public servants. My grandfather, parents, uncles, and now brother and brother-in-law were and are all cops. My aunt was a 911 dispatcher, and my grandmother was a nurse. It is a family legacy that is deeply ingrained into who we are and how I was raised. My parents never pressured me to join the family "business", but now that I haven't, it's hard not to feel like an outsider among the people that I'm closest to.

My parents and my brother have a shared experience and connection with each other that I will never have. And it is hard for them to truly understand my work and all that it entails. I'm lonely when I'm with them, and I'm lonely when I'm away from them. They wonder why I cling to my phone, the gateway to every connection I have with all of those long-distance friends and social media groups that I lean on for understanding.

From the moment I graduated college, I applied the policy of walking through whatever doors opened to me. I take the job that I'm offered. That's how I ended up in North Dakota, where I will soon be returning for my third season. The LC article talked about good and bad supervisors. I may not know where I am or where I'm going, but I do know that I wouldn't have gotten this far or be the person I am today without the mentors I've met during my brief time in this career. I feel very lucky in that respect. One of the few reasons that I am going back to North Dakota is for the supervisor who has guided me and taught me so much over the last two summers. I owe him a lot and will be very sad when the time inevitably comes for me to add him to the list of people I'll probably never see again.

I still feel stuck, stagnant and uncertain about my future. I am considering a master's degree because I feel like I have to, even though I have no desire to live on a college campus for another two years or put myself into another $25,000 (USD) of debt. I am considering a career change, simply because the longer I do this, the more I crave stability and a physically present community instead of just a virtual one.

I'm tired of feeling like a failure and ashamed of barely earning enough money every year to support myself. Of still being single because I'm never in one place long enough to establish a relationship. Of feeling emotionally exhausted from the constant cycle of goodbyes to people I care about. This all isn't to say that I don't feel incredibly blessed for the people I've met, the experiences I've had, the places I've seen and lived, and the stories I've accumulated. But I'm not sure that's enough.

I don't know how to find my passion in conservation again. I question what the point is to the work that I do and how much longer I can do it before I completely lose sight of the bigger picture. Even the guidance of my mentors, friends, and this beautiful new community hasn't given me the clarity that I feel like I need to move forward.

# | 4 |

# Fearing what I love

## RENUKA KULKARNI

*"Oh, you're a wildlife person? But don't you sit in a library all day?"*

We all have that one aunt/uncle/extended-family-member-you-don't-remember who can sniff out weaknesses like a bloodhound, and who just never wants to try to understand what you do for a living. *"I'm doing a PhD"* is a somewhat acceptable answer, but you're also praying inside that they don't find out about your lack of funding.

Yes, you Indian aunty, my personal abyss agrees with you that I am a burden on my parents: a 27-year-old, unmarried secretly depressed burden who watches MasterChef Australia (MCAU) reruns to get through days paralysed by self-loathing. I'm Renuka, by the way. Doing a PhD in environmental history, trying to highlight the problem of invasive plants wrecking unique habitats in India, by combining evidence from vegetation surveys, historical natural history accounts, and current conservation programs. If this sounds like a reach, I know it is. My PhD thesis was born out of my experiences as a chemical ecology intern, an archivist, a science writer, and a realisation of how decisions by various people throughout history shape our natural landscapes.

My field site is one of the last evergreen forest patches in India: beautiful, mountainous, quiet in some spots and crowded with tourists in others. I love it there, but there is a part of me that fears those dense forests, and that has been a bitter realisation. Most conservation biologists do what they do because they love the natural landscapes they roam, and they feel at home in those quiet woods/waters/mountains/swamps, away from the crowds. At least, that's how it appears to me. Whenever on field trips with my classmates or volunteers, I always admired how confident they were during vegetation surveys or their ease in identifying birds or butterflies or reptiles.

Growing up, I was a fairly confident kid with good grades and a lot of extracurricular activities, but not a lot of friends. As I've grown older and chosen careers that seemed fascinating and great opportunities to learn, it has also isolated me more and more, feeding into a lack of self-belief, feeding into the fear of people and their judgement, to the point where I feel a complete absence of a 'self' at times.

You'd think I'd feel at home in a dense forest with trees that don't ask questions and creatures that don't expect you to socialize with them. But I don't. I feel afraid of those quiet, lonely, sunlit places, or maybe, it is my PhD that has made me feel this way. It is perhaps my PhD, struck dead by a pandemic, unfunded and geographically isolated, not quite completely understood by anyone except me, and I am paralysed by being an archivist in a wildlife conservationists' world and a wildlife conservationist in a historians' and political scientists' world. This comes after being a biologist in a journalism school, a biodiversity student in a newspaper office, and a journalist in a classroom of wildlife biologists.

*So* many labels, eh? I can't decide if these words are a form of narcissism, an arsenal against my imposter syndrome, or phrases I've heard from loved ones trying to bolster themselves that I am worth all their efforts and sacrifices.

Perhaps it wouldn't have been that hard, had I only been somebody else, I sometimes think. If only I was someone who could talk more and didn't always castigate herself for being too boring and bookish. I've always loved books. I've always felt consumed by them, and I always read the same books, again and again, a million times because I find something new in them every single time. A part of my thesis data collection involves looking through old herbarium records and government correspondence, and I love going through those neglected, often mouldy and dusty pages stored in old-fashioned cupboards. But even there, fear has crept in.

Is it the PhD that I fear? Built on a lack of trust about my abilities, amplified by the process of this enormous, unique professional experience that is famously said to teach you more about what you don't know? It makes me afraid of missing out on a crucial record. It makes me afraid of this lovely landscape I've chosen to study. My mind keeps running to the quiet seaside projects I once volunteered for, and I sometimes wish I'd chosen marine biology as a career. But then again, if I had, would I have started to fear the sea as well?

Fear. It lurks in all of us, accompanied by despair, followed by a sense of loneliness. Fear is what keeps an organism safe from harm so that it can survive and live to see the day it can procreate. But fear also keeps you in your comfort zone, and no episode of *MasterChef* goes by without the phrase:

*"You grow far more when you're out of your comfort zone than when you're in it."*

I should know: I'm watching MCAU season 7 right now even as I write this, and Reynold, the dessert king, is trying to cook fish perfectly, something he's reminding us is out of his comfort zone. (He loses this challenge, by the way. And gets eliminated.) Some people drink to numb their insecurities and fear. I watch MCAU reruns, episodes I have already watched before, because I know how they end, and I am reassured by that, by the knowledge that I know what's going to happen.

## The root of fear

I live with my parents and my cat. I don't have a lot of friends, because most people my age are married with full-time jobs, and are too busy to spare much time. My sense of guilt stems from the fact that I do not earn any money, or at least, not enough to be financially self-reliant. There are intricacies of this that I wish I could write, but I can feel the fear rising in my chest at the thought of it. At the thought of writing things down for you to read, and then judge me for it. To an outsider, I appear absolutely privileged, and I am. I have a roof over my head, parents supporting me, good and kind PhD supervisors, a loving and encouraging boyfriend, a brat of a cat who makes me laugh every day. And I'm still unhappy. I love my subject and I'm excited about getting my PhD before I'm 30, but the fear, that black, suffocating Dementor, keeps hovering.

The Dementors keeps coming: *you're out of your depth, you'll get attacked by bison and leopards during your fieldwork, you don't know your plants as much as you think you do, you should know your archives through and through, your work is not going to make any difference whatsoever.* But most of all, the biggest Dementor of all: *you're a fraud.*

My Patronus: *I just want to do the best I can.*

The only way out of this is through it, and my mind knows it. So, I'll fight it and go through it every day. I'll fight the imposter syndrome by checking and rechecking my references. I'll fight the sense of failure by doing something that makes me feel positive, like learning how to drive a car. I'll fight the fear of thick forests by walking through them, looking for familiar plants. I'll not fight the wild bison and leopards, but just keep an eye out for them and not get in their way.

I'll just try and do the best I can. Every day. Because that is all that I, or anybody else, can do. And I hope that the person reading this takes a deep breath, and hopefully feels a little bit stronger to live out another day as a conservationist.

## Update: August 2021

Some years, even the worst of them, can still have their moments of sunshine. Your PhD can flicker out and die, making you feel more like a waste of space than you've ever felt, the world outside burns and dies, and you'll sit outside a hospital's operation theatre as your father undergoes a life-changing surgery and become aware of the torturous side of time.

At the same time, you find people who give you a lifeline and a way back into careers you thought you'd have to let go of, your boyfriend sits beside you in that hospital and asks you to marry him, and your drowned research ideas get resurrected.

Invasive species is still my focus, their history with people and land-scapes is still my research goal. I am sad to leave my beautiful field site behind, but looking forward to studying the urban history of invasive plants. The world outside is still on fire, and getting worse by the day. I guess you just have to push on, always push on (an empowering mantra of another MCAU season 7 contestant), and try to find ways to make things better, one day at a time.

# | 5 |

# From refuge to responsibility

## PHIL MCNAMARA

It's ironic that a child, who'd come to love spending time in nature as a way to avoid conflict at home, should be forced to confront that fear of conflict because he grew up to work in the conflict-ridden world that is conservation. When I was very young, my family spent three years in an outback mining town in central-west Queensland. I have wonderful memories of the feel of the remote natural places we visited in our regular travels out of town. The photographs we have of this time are of a happy and connected family.

But when we returned to South Australia without my father, we were all broken. I don't think my mother, sisters and brothers would mind me saying that the event of my mother leaving our father drove a wedge between us. Conflict seemed to be everywhere for me during those years and I'd often be found alone in our shed or heading out on foot to remote natural places in the Adelaide Hills. They were my places of refuge. And I never had a bad experience out there – going into the wild.

I still get that same feeling today when I'm alone in the forests and woodlands around my home, or the mallee (a type of scrubland vegetation common in southern Australia) at work, where my mind and body just relax and all my worries vanish.

I'm not sure if it was the circumstances of my growing up or if it was simply in my disposition, but I emerged from my childhood with four things: a love of natural landscapes, a strong sense of environmental and social justice, a desire to fix wrongs but also, and unfortunately for these other characteristics, a fear of conflict. Given this last characteristic, it might seem unusual that my first 'career' was as a police officer.

It didn't take long for me to realise that I needed to get out of that line of work, so in my spare time, I started a science degree in conservation and parks management. This was one of the best things that I could have done because it led me to work with some of the kindest and most passionate people I have met.

Something else that was formative for me as a young adult was that I started reading books by Charles Darwin. In particular, I loved *The Origin of Species* and *The Voyage of the Beagle*. I say they were formative because growing up, I didn't feel like I had any kind of belief system instilled in me, and I had many questions about life. When I read *The Origin of Species*, it made sense of the world to me and became my belief system.

It was not until I started my degree that I began to truly understand the impacts that humans have, and are having, on our climate and biodiversity. One of the most important realisations I had was the speed at which humans are now changing this planet, quicker than any other time in human history. This makes me sad. True to one of my characteristics - a desire to fix wrongs- I finished my degree and from there, my work as a conservationist spiralled fantastically out of control.

At first, it was not in an advocacy sense; fear of conflict got in the way of that. My first job out of policing was as a Green Corps supervisor, working with a team of young adults on re-vegetation projects that created habitat for threatened species like the glossy-black cockatoo (*Calyptorhynchus lathami*). From there, I volunteered in bushland management all over the place and have worked as an urban conservationist, bushland manager with the National Trust of South Australia and project officer in natural resources management across the South Australian Murray-Darling Basin.

During my time with the National Trust, another formative event happened in my life: my daughter Darcey was born. It made me realise, for her sake, that I had to be more outspoken about how we treat this planet of ours. It also made me want to instil a love of nature in Darcey so that she could be an advocate for change herself. In hindsight, this was a selfish expectation that deflected my own responsibilities to be outspoken. When she was born, I wrote this poem to her, which is a good reflection of my thinking at the time:

> Legacy to Darcey
> *I often think about when I die,*
> *what I leave to you:*
> *two cars, a house, some money, a debt,*
> *three cutlery sets, books and files,*
> *crooked sky and oily seas,*
> *raped earth and plastic fill.*
> *But of beauty, change and substance,*
> *my legacy is you.*

It was about the time I wrote this poem that I decided to write a novel as my way of being outspoken. Engaging in conversation with a climate change denier or someone proposing to clear bushland was not something I did very well. I didn't have the confidence to be calm and articulate on those occasions. I get the sweats just thinking about it. Conflict.

So writing enabled me to be articulate and formulate an argument in my own time. It's another one of the best things I could have ever done. I have just self-published my first novel, titled *Red Reflection*, which is essentially about habitat destruction and how quickly we have changed the world. It's also about what it feels like to sit on the outside of popular belief. It starts and ends on a sad note but my sequel, *Red Hope*, which I've started drafting, has a more positive beginning and end. This one is about climate change, and I can only hope that the ending reflects what we can do over the next decade for the sake of our children and their children's children.

My latest ventures are plogging and a small business called The Third Fuse Project. The word 'fuse' is a contranym, which is a word that has two opposite or contradictory meanings. Fuse can be a device to detonate an explosive charge (Fuse #1: climate change; and Fuse #2: habitat loss) or it can mean to unite or blend into a whole (Fuse #3: people and nature coming together).

The Third Fuse Project is essentially me being a writer for the environment, focusing on writing about three major Fuse #3 principles: living a modest life (living sustainably and lowering our carbon footprint), communicating environmental science through art, and respecting nature in everything we do and everywhere we go. The other venture, plogging, is where joggers pick up rubbish along the way for the sake of the environment. I do lots of trail running and allocate one run a week to plogging and have done over 100 plogs, which is a lot of rubbish off our streets and out of our creeks and parks.

I am proud of what I have done as a conservationist but feel like I need to take another step forward and be more willing to have face to face discussions with my family and others about our two biggest world emergencies (or fuses): climate change and habitat loss. It's great to be part of *Lonely Conservationists* because it makes me feel like I'm not alone in those difficult conversations.

# | 6 |

# The journey of a wildlife enthusiast

## NISHAND VENUGOPAL

It all began with an interest in photography. I always loved taking photographs of wildlife and nature. But I didn't want this passion to be restricted to just being a hobby. I felt I could do much more.

I was born in an era that saw the explosion of technology. The pace at which technology has developed has been so fast, but thankfully, I was able to keep up with it. Born in 1979 and known as the Xennials who precede the Millennials, our generation has seen an analogue life in our childhood which turned into a digital mode when we reached adulthood. This also means that we have seen how the Earth has changed during all these years—the good and the bad.

A major part of what constitutes 'the bad' is occupied by the devastation caused to our environment. Our cities are highly polluted, our green spaces are being taken over by the real estate industry and the natural spaces that are inhabited by animals and birds are in grave danger. We have forgotten how important it is to preserve nature. Many of us are not even aware of the blessings bestowed by nature. I want nature to be preserved so that the coming generations can enjoy the benefits of it, the way we have.

And that is why I believe it is my duty too to serve the cause of conservation in my own way, through creating awareness and using the power of social media for it. As Dr Jane Goodall wrote:

*"What you do makes a difference, and you have to decide what kind of difference you want to make."*

My duty as a wildlife enthusiast is to learn about the value of biodiversity in India. That means travelling and reading about it. I am grateful that there are a lot of books that throw light on this rich subject. The more I read, the more interested I became to share information with the world, especially with kids. Kids tend to naturally have a bond with nature – they love to play in mud and water. We have experienced and witnessed this quite commonly in our childhood.

Today, most children are hooked to gadgets and prefer to remain indoors. The air pollution in New Delhi, where I live, is quite high; I always feel anxious whenever my kid plays outdoors, wondering what noxious gasses would affect her. The worryingly high level of air pollution has pushed people to think and work towards preserving nature. To add my bit to such efforts, I volunteered with WWF India for their movie-making project and then their program of taking kids on nature trails. I am also writing for a children's newsletter — my photographs are also helping in sharing my thoughts with them.

I have travelled through various terrains in India – from Ladakh in the Himalayas to the coast of Odisha – to see and experience the flora and fauna that grow there. Only when you see nature and wildlife in their original form, will you learn to respect them. There's a story in each photo I take and every person I meet during these journeys, and these are what I share through my social media channels.

When you are with nature, you get to learn from other species too. I witnessed the virtue of the motto, *'Keep on trying till you succeed'*, from the tiny Olive Ridley turtle hatchlings on Odisha's coastline. They move towards the sea but are thrown back into the beach umpteen times by the waves, and yet they don't give up. They work hard and finally reach the safety of the water. The sight itself was a revelation to me. It was a valuable lesson I learnt from nature – I, who had taken a leap of faith to help it.

For the past year, after leaving my job, I have been observing and documenting conservation works happening in India. I have great respect for young wildlife biologists and conservationists who are working hard to preserve the precious flora and fauna of India. I made up my mind to ensure their efforts reach a wider audience and let people know that there is something of worth left to love and care for. From discussing and volunteering with caretakers of sloth bears and elephants in Wildlife SOS care facilities to learning from the researchers of Wildlife Institute of India about the latest techniques they use in conservation and forest management, I have experienced a huge learning curve.

I got to know how people who live near the forest area try to mitigate human-animal conflict. I also got to know about and meet members of an organisation named Kumaon Matti which is trying to face such conflicts with a positive attitude. I am glad there are a lot of people who are working round the clock to preserve the rich natural resources of India. Generating interest in biodiversity and the importance of it in providing us basic necessities of life like clean air, water and healthy food is my mission. People need to know that all is not lost but that yes, everyone has to do their bit as time is ticking. It is now or never, but sharing stories of hope will definitely inspire more people to work towards saving nature.

A healthy nation needs to preserve its natural resources. That is our real wealth and the legacy that we need to pass on to future generations. Hope is still there and I am optimistic that a change in our approach is possible and things can be turned around to make Earth a better place.

# | 7 |

# My hometown is on fire

## JESSICA LECK

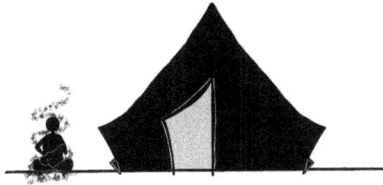

My name is Jessica Leck, and as I write this, my hometown is on fire.

I live on the mid-north coast of Australia, and this week has seen an area roughly seven times greater than the Amazon burn across my home state. Friends have lost homes and others have lost loved ones, and still, my community and their government will not speak about the elephant in this room.

But to understand my thoughts as I watch the flames through the window of this evacuation centre, you should know my back story. The mid-north coast has not always been my home. Until earlier this year, I was one of the inner-city greenies being criticised by the politicians.

There has never been any deviation for me: I was always going to be a conservation biologist. With a family unable to help and a complete lack of funding in this industry, this need of mine was never going to be easy, but it was always going to be the goal. With the support of a wonderful partner and a work ethic that fortunately placed me in positions with roster control, I was able to work full time to support myself throughout eight years of university; two degrees, a graduate certificate, countless internships and a traineeship. If you read that and think that may have been overkill, you would be correct, but you also probably have not tried to get a paid graduate position in conservation in Australia.

My twenties have not been an easy period of my life, but my never-failing drive to do everything I can to preserve the wonder I fell in love with has always been the only motivation I needed to get through a 20-hour day.

That brings us to early 2019; two graduate positions under my belt, I considered myself unbelievably lucky to have gotten a third, as a junior ecologist for a small consulting firm. The 70-hour weeks, an hour-and-a-half commutes and AUD 60,000 salary were supposedly everything I had been working for, and yet there I sat in a high rise, signing off EIA's for clearing extensions to landfill sites. It was supposed to be the dream, and all my university cohorts were hitting me up for introductions, but clearly, my subconscious knew something I wasn't ready to admit. Two months in, my partner came home to tell me he had gotten a new work placement three hours north and had to be there in three weeks, and despite this recent fulfilment of the "dream", there was no hesitation as I sat down to type out my resignation letter.

So three weeks on and there I was, the most highly educated, overly qualified housewife on the mid-north coast. The relief offered by the reduced cost of living meant my income wasn't necessary, so I spent the next few months fulfilling the dream in more practical ways, my only restriction being the heavily right-wing mentality of my new community.

I was resigned to this life, knowing my opportunities were vastly limited, my opinions largely unwelcome and finally convinced of my limited capacity after years of being told I wasn't good enough by Sydney. So imagine my surprise when I was offered work lecturing at the local university, and then again with a local connection. This community I assumed to be closed-minded and archaic had offered me more opportunities to do good than Sydney ever had.

So this was when my tumultuous love affair with this community peaked. A more diluted distribution of conservationists means we have the opportunities and funding to positively contribute, yet much of the community continues to see my only value being in helping them increase the productivity of their over-harvested soils. The words 'carbon' and 'biodiversity', have very limited application in their minds, and single species don't dare merit a mention. But still, public lands and the support of the few converted have allowed me to do more good in these three months than I have over four years in the industry in Sydney. I have saved countless trees through conversations with farmers and spent hours standing in front of their children teaching them the interconnected value of eco-literacy.

And yet, I sit here in an evacuation centre while our home burns around us, not listening to discussions of climate scenarios and mitigation. Instead, my ears are flooded with the beginnings of terrifying irresponsible ecological demands; clear-felling borders and grazing national parks, while further in the distance the anti-greens mentality is being further cemented with misinformation.

I know every tree I have planted and every shrub I have preserved are now charcoal, and hold a deep fear they will be replaced with the security of grazing land. The ecological value I had spent these months communicating to these landholders represents nothing but fire risk to them now. In a region where the far right-wing conservatives run unopposed every term, I am finding there is only room for my ideas when all other variables are able to remain the same. I wonder, is this mentality regression an aspect of the positive feedback cycle missed by some scientists?

So what now? Well, of course, I keep going here in this closed-minded community. Where my ideas are unwanted but have the room to grow and inspire. I go to work on Monday and I teach my students about the importance of megafauna in nutrient cycling and then I host my carbon sequestration workshop for farmers that night. Because there never has been and there never will be another option for me, I am a conservation biologist, whether I am employed or not, lonely or in front of an audience; this is the only option for me.

# | 8 |

# Jungle lass: Wildlife biologist turned media pro

## LYNETTE PLENDERLEITH

I've heard that if you have a Plan B, you are not fully committed to Plan A and you might as well wave it goodbye. Maybe that's right. I had both a Plan A and Plan B and I worked exclusively through Plan B for well over ten years. Maybe I've sidestepped the inevitable only by doing both now.

I grew up in a remote area. Not going-to-boarding-school-because-there's-no-other-option remote, but no-tofu-in-the-supermarket re-mote. There wasn't even really a supermarket until I was a teenager. The supermarket was roughly what would now pass as a large "Express" version you might find at a fuel station. But this piece isn't about supermarkets, despite current appearances. This article is more about the forests and the fields within a few hundred metres of said supermarket in my hometown. It's about the beautiful rolling hills of England that were my home for the first twenty years of my life.

Going for a walk as a child showed me rare orchids and badger setts, birds of prey and ancient fossils, third-order streams and acid grassland ecosystems. My family were all biologists of some description, with an emotional connection to the country and the knowledge of all its inhabitants – be them slimy, scaly, feathered or furry.

It hadn't even occurred to me that conservationists were necessary. Surely everyone with the ability to see, hear, and feel would just look after the things that needed protection?

I wanted to work in film and television. I used to lie in bed at night pretending to review my books for an imaginary camera suspended from my ceiling. I wanted to be a presenter. I enjoyed drama and acting, singing and dancing. But I was advised against it. It was a silly idea for a girl from the country. Perhaps I'd like to be an English teacher?

No. I didn't want to be an English teacher. But maybe I could be persuaded to do something outdoors. Something that required a waterproof hat and a pair of gumboots. I liked animals too – a lot, especially wildlife. And being out amongst it, in the rain, in the sun, all year round. I didn't love the cold so much, but I could tolerate it for the bird-watching. A love of the land runs through my blood, and my upbringing amongst the trees and the weather and the hills gave me a sense of responsibility for the wildlife and the habitat in which it lives.

As I grew older, the need for conservationists became more apparent to me. Climate change registered on my radar, along with endangered species and habitat destruction. I went off to university to study biology and environmental science, with the aim that I will make a difference. Then one day, in the rainforests of Indonesia, I met a frog biologist. Now that sounded like a nice career – outdoors with the wildlife, a good bit of writing, and frogs don't like the cold either! So there was my Plan B: go to the jungle, study frogs.

I'm not sure it was the alternative that those people had in mind when they told me to think of something other than media.

I had hoped to do a Master's in Wildlife Management and Conservation, but my grades weren't good enough for the course I wanted to do. I also had very little work experience and couldn't even get an interview for the kinds of wildlife biologist jobs that were available in the UK. So, I started sending speculative letters around the world to anyone that might have a job that would interest me.

I got really lucky: I got offered my first job at the Amphibian Research and Monitoring Initiative in Maryland, USA. That was easier than I thought. Less than a year out of university and I had already made it! It was only a three-month contract (the length of the frog's breeding season), but that first job opened doors for me all over the world.

A dozen years, one Master's degree and a PhD later, I was still not only hankering after the glamour of a media career, but I had also come to recognise the value of conservation through storytelling. The accessibility of television and film, in particular, makes them amazing outreach tools. And although the nature of television is changing, the popularity of the screen medium continues to grow.

And I got really lucky for a second time. When I was a PhD student, I sat on an advisory board for a sustainability course; there, I met a TV producer who was looking for someone to fact-check a children's wildlife show. The rest, as they say, is history.

I work freelance now as a science media professional. I present, write, and research science stories. I have my own film in pre-production and I'm producing a podcast. I founded *Frogs Victoria*, a state-wide amphibian interest group where I really get to flex my outreach muscles, and I am President of the Victorian chapter of Australian Science Communicators. I even continue to feed my herpetology addiction by occasionally working on short or part-time science contracts. I suppose you might call it Plan C.

The freelance life can be a lonely one. Always the outsider, often the leader, the loneliest positions of all. In the ecology world, most people that I meet are working towards the same aims, even if they aren't on the same project. But in the media, there are people with every imaginable stance on conservation. Even when we're lonely, we conservationists are not alone because we stand on the shoulders of giants. We are not alone, because we come from a rich heritage of people that made us care. Whether because they didn't care enough or couldn't do something about it, or whether they cared a lot and worked hard to make other people care. We were all brought up by people who have given us a reason to stand up and be counted as a global team of conservationists, scientists, teachers, and activists.

# | 9 |

# The perks and pitfalls of a never-ending conservation obsession

## DAVID DE ANGELIS

The word *obsession* gets thrown around a bit, but maybe understandably, true obsession seems to frighten most people. Conscious of the other syndrome that many lonely conservationists have talked about *(impostor syndrome)*, I still feel the need to point out the difference

between having an obsession with natural history and necessarily having expertise in ecology or wildlife conservation!

Few people seem to have clear memories of their time in kinder-
garten, but some of mine are still vivid. Social introversion and an
OCD-like repulsion of human 'mess' kept me from regularly interacting
with more than a couple of the other children. Yet, I was the only one
who constantly begged to play with the pet rabbits and guinea pigs. A
lot of my time was also happily spent in the company of a sulphur-
crested cockatoo and an Eastern long-necked turtle. Visits to a child
psychologist ended soon after I told him that I coloured a drawing of
the Yarra River brown rather than blue because of sedimentation!

Despite my challenging childhood obsession with nature, I was
never taken on anything more than day trips within a couple of hours'
drive from Melbourne. Camping and holidays weren't on the family
agenda, so most of my early natural history learning came from books.
Apart from the Gould League and Steve Parish, most of my available
reading was on northern hemisphere wildlife, resulting in familiarity
with the North American copperhead (a pit viper) years before learning
that we have copperheads belonging to an entirely different family of
snakes in Australia.

It also led to a constant turnover of unusual pets. I was scared of
dogs, so over time I maintained the company of a cockatiel, budgies,
finches, quail, geckos, blue-tongue lizards, bearded dragons, frogs, ax-
olotls, fish, hermit crabs, snails, ants, spiders and stick insects.

It wasn't a coincidence that the secondary school I went to had a
voluntary student environmental group, offered a middle school environmental management subject and had an area of bushland adjoining
a creek. The usual teenage distractions of sport, clubbing and gaming
never got a look-in. Maybe unsurprisingly, I spent many lunchtimes
and some of my other spare time either volunteering in the school's environment centre or down in the bush. This led to my first job, working with an environmental land management company at the end of my
final year. In fact, I have never worked outside the environmental sector, which seems incredibly fortunate or even envious to some people,

but it isn't without some regret as it has meant my broader life experience has been further limited.

Despite struggling with maths and other analytical subjects, enrolling in biological science at university seemed inevitable. Although I found completing my undergraduate degree far from easy, it gave me an irreversible appreciation for experimental design, critical thinking and undertaking objective research. Even so, when scoping potential projects for an honour's year with one of my lecturers, he commented that I should have probably been born 20 years earlier. I thought 200 years would have been more appropriate!

Even in my final year at university, when I was studying the ecology of burrowing skinks in semi-arid South Australia, opportunities to volunteer on other projects distracted me from what I should have been focused on. Nevertheless, I entered the ecological consulting industry straight after graduating.

I have remained in the same job ever since, although I am considering returning to study. Regardless, I have never had the usual lifetime desires of owning a home or anything more than a functional car, simply because I would much rather invest most of my time and savings indulging in wildlife conservation and natural history.

While the few of us who truly have an obsession will never really change, our challenge is to make sure it can be used constructively for the benefit of conservation, and not undesirably impact the people around us.

I'm indebted to many colleagues, family members and like-minded friends who have not only been incredibly tolerant but supportive of my journey so far. As the words 'lonely conservationists' suggest though, many of us could probably do with a little more social interaction. Some of us who are based in Victoria and have contributed to this book occasionally catch up at environmental seminars, conferences and other events. Maybe there could also be room for more regular social events under the banner of *Lonely Conservationists*?

# | 10 |

# Sea turtle researcher or housemaid?

## LISA

*All names in this piece have been altered to protect the identities of the people they are about. Names are being used for clarity purposes only.*

My therapist says that there are only a handful of decisions that people make which will truly impact the rest of their lives. For most of mine, every little decision felt like I was breathing with a brick on my chest.

I was the kid who would only take *one* piece from an unguarded bowl of candy on Halloween. I am the adult who returns library books early, not because I finished them, but because I am nervous about the librarian confronting me for $.03. Too dramatic? Probably. But why do you think I'm in therapy?

I remember accepting my job offer as a sea turtle biologist for the 2020 season. I was jumping with joy over my first paid position working with sea turtles. Sea turtle conservation became my passion since a memorable childhood vacation when I was asked to help relocate wild nests as a hurricane approached my hotel in Mexico.

I can't tell you where it exactly went wrong. But it went sour, fast. I started off working mostly in the office: sending out turtle patrollers, responding as dispatch, tending to equipment, and being a field lead when needed. I thought, *"typical beginning of the season slow work until the animals start nesting".* I looked forward to the end of the season when I was promised to work in our incubation lab and do the sea turtle hatchling releases.

What was described to me as a six-month stint of working on a famous conservation project, turned into a janitor job. I do want to give one of my bosses, Emily, credit where it is due. In the beginning, Emily tried very hard to get me into the field. She knew this was my second time applying to work for this organization and how badly I wanted to be there. However, it was not enough to make up for what was yet to come.

Instead of working on papers, probing for nests, doing necropsies, or even interacting on the social media page, I was given a list of chores every day. I went into the giant dumpster and recycling bin to move garbage around. I cleaned the toilets. I sanitized the necropsy lab. I washed cars and UTVs for people who were in the field. I filled up people's gas. I cleaned our office kitchen. I broke my back moving ice chests for people camping. I reorganized and took inventory of every single closet on their property. I became the person who was assigned all of the office chores while co-workers were spending time camping, doing GIS work, or on the organization's social media pages. Even co-workers hired to a lower position than I did not have half of the janitorial duties I did.

I took matters into my own hands multiple times and inquired about different projects I could work on. Every time I asked for an assignment, I was told to sweep. Every time I asked to edit a paper, I was told to empty the dehumidifier. I was feeling frustrated but kept telling myself that this was a part of being in conservation.

Until one day, I finally got picked to help design a logo for our department. I looked forward to it every day. I was excited to be working so closely with the head of our department and my personal hero, Ashley. I thought this would be my chance to prove my worth above my cleaning skills and get recognized. I had dozens of designs to choose from. I even got my friend, a professional artist, to draw us a sea turtle *for free* to use. I worked on them in my free time just hoping to contribute to the legacy of the organisation I was serving under. Ashley eventually stopped answering me, or even acknowledging me in the office. This stung colder than usual because not only had she been my hero for years, but my desk was directly outside of her office.

I also learnt that later on, she did not know my name. She didn't know I was the person she was emailing. I was one out of five people who worked in the office.

As time went on in the office, a co-worker (let's call him Dave) started spending more time chatting with me. Things started escalating from a "*how are you*" to him explaining his sexual life with his wife and her fantasies of having a threesome with another woman. This was followed by a request from Dave on Facebook, comments about how his wife finds me attractive, and stories about his genitals. Unfortunately, Dave and I had to work closely together in a team with two other men. Dave thought it was appropriate to make me the end of a few sexual jokes which, thank God, my other male co-workers did not approve of. Clearly uncomfortable and completely outraged, I had to speak to Emily about it and be moved to another shift. I did not want him fired, but I wanted to be comfortable at work. Over the next month, I found Dave speaking to our female interns about his showers at work after being in the field. I made it a point to interrupt the conversation but he completely ignored me. At this point, I made one of my male co-workers come outside to interrupt. I stayed outside with those girls for a little while after. I refused to let him abuse the power of his uniform over interns.

I kept doing my tasks and the days kept getting worse. My feet dragged across the floor with the thought of doing more chores and being berated for the ones I didn't do well enough. A fire grew inside of my chest every time my shift overlapped with Dave and he made a sexual joke or tried to explain things to me that I already knew. If he patronizingly explained to me that tides go up and down (when I was a recent graduate student in marine conservation) one more time, I was going to lose it. The only reason I didn't snap a keyboard in half is because Emily told me that I was going to be in charge of the incubation lab. FINALLY, this was my break!

Oh, how wrong I was.

I spent my first two or three weeks working from 7:00 pm - 3:30 am alone. Which would have been fine but our office is isolated 20 minutes from any town with no other staff was on duty, and I am a 5 ft tall, 125-pound woman. On my first night on the job, our office administrator pulled me aside to say I needed to lock all of the doors and never go in the parking lot (where my car was) because a pack of coyotes hung out there at night. And on the first night, she and her husband stayed with me for an hour after their shift because they were so concerned that I was there alone. I sucked it up for those weeks, fought through my exhaustion, and battled the fear of some drunk beach camper trying to break in.

I finally had two people come to stay with me overnight. One was Ashley and the other was a volunteer of 11 years, Maureen. I was so excited to learn from them. I had to spend the first two or three nights of that week in the necropsy lab alone with no bathroom because - according to Maureen - Ashley wanted to sleep and we made too much noise. The only way to access the bathroom was by walking near the parking lot with the coyotes. Maureen was only allowed to start sleeping on a cot in the office after nine years of night work with Ashley.

Eventually, Emily told them that it was borderline illegal for me to be in those conditions so I was allowed to stay in the general office. Whenever Ashley emerged from her cave, I was never greeted. It was always *"sweep the floor, it feels sandy"*. There is no closer way to be to your boss than seeing them in their pyjamas at 2 am going to pee, but she still didn't care to learn my name. Once again, due to the fear I was being overdramatic, I said nothing.

As the turtles started to hatch, more people stayed overnight with me. At first, the volunteers and another experienced biotech, Gina, were being sent out to release them. Which was fine: the volunteers need to be recognized for their free hard work. It became a problem when I finished all my janitorial duties and was still not allowed to walk down the block to see the volunteers release sea turtles. It became an even bigger problem when I asked Emily and Maureen to release the hatchlings and she sent the interns and co-workers who were not on any overnight shifts.

What really sent me over the edge was when another co-worker, Adam, the same supervisory level as me, got to release two nests without doing even a single night shift. I had been on the night shift for at least a month and a half at that point. I put in the effort in the incubation room and the office, I should have gotten to send at least one nest off.

My mother and sister had to come to visit me at this point because I was so distressed. I called her every day crying. I dreaded going to work. I craved being in my bed. I was isolated from a lot of my friends due to our conflicting schedules. I was across the country during COVID where people in my life were getting sick and dying. I looked into the mirror and noticed I was changing. What happened to the big, genuine smile I usually wore? Where was that sense of humour that always could cheer other people up? That person was holding on by her fingernails but she was about to freefall. I told my mother and sister that if I went to a hatching just once, I would feel validated and could push through the season. They said, *"why don't you just go?"* I responded in typical goody-two-shoes fashion, *"My bosses said no."*

That same night, my co-worker, Gina, and I were in charge of the lab and a nest was ready to be released. When sea turtles start to frenzied, they must be released immediately. They will start to scratch at the containers we keep them in and that noise can cause other nests in the incubation room to hatch prematurely. Gina was already coming back from releasing turtles with a first-time volunteer named Debbie. I called my housemate and he agreed to drive my mom and sister to the park. I spoke to the volunteers, the Smiths, and asked if they would mind if my mom and sister came to watch. The Smiths had no problem and offered for them to drive my family in the government vehicle. Me, knowing that wasn't allowed, thanked them for their generosity but said they would be taking my car.

I tried to pawn my keys off on my family and housemate to go watch the hatchling release. But they refused to go without me. This was a once-in-a-lifetime experience for them, and maybe, for me too. I checked on all the hatchlings before I left, took the proper safety precautions, and radioed Gina who was just 10 minutes from the office. I kept the door unlocked for her because I was told that I was the only biotech who had a key to get inside. I left the door open for her because the nests had to be checked on every hour and sprayed with water. There were even more volunteers in the trailer outside our building that kept an eye on anyone entering the building. My mom and sister used the restroom while the Smiths came inside to help me put the eggs in the truck for transport.

I went on the hatchling release. I knew other biotechs were going when they weren't scheduled but I was too much of a rule-follower before this. I was tired of being overlooked and left to clean the toilet. I may have felt like I was going to vomit the entire time but I got one of the most beautiful experiences of my entire life with two people who mean the most to me. I remember pulling up to a dark stretch of isolated beach where the city lights couldn't reach. The moon perfectly lit up the path for the little ones to settle into the docile waves of America's Gulf Coast. My sister, who is terrified of animals, even built up the courage to help move a hatchling out of its crate. It was a moment for my family and myself to realise why I went through all of the schoolings, all the days in the field, all the late nights of studying, injuries, bug bites, and even the bullshit of this job. I did so much because of these moments where I could save the wildlife that is so much more important than myself.

When we pulled back into the office, I heard the fire alarm go off. I ran directly inside to check on the people and turtles. It turns out that the building had a faulty fire alarm and none of the night employees was told how to shut it off. However, I was the one who figured out how to. Debbie, Gina's volunteer, told Emily about the entire venture and then claimed I left work in the middle of the night. However, I stayed at work for my entire shift and had employees who were on overlapping shifts that could vouch for me.

Emily accused me of putting the lab in danger, although I followed all the safety procedures. I was accused of letting my family in the incubation room when the truth was that my family was in the bathroom (only certain personnel were allowed in). She accused me of being irresponsible because the turtles could have hatched prematurely with the fire alarm. Which, is untrue. Sea turtles can be sparked to hatch by low-frequency noises such as a man's voice, not a high-frequency sound like a fire alarm. I was accused of leaving work and going home.

I quit.

However, I felt guilty and wanted to try to make amends. I know it technically wasn't the right thing to do and no one will be harder on me than myself for breaking a rule. I respect Emily still to this day. I told her the way I felt and she told me that she felt like she let me down.

She did. She broke my heart. The result of this conversation was me not quitting but being put on the day shift as punishment.

I technically broke the rules. I completely own up to it. I was just so desperate to hold onto something. I wanted to keep the facade that I had of working for my dream organisation, doing my dream job and working underneath one of my personal heroes, Ashley. I stayed for two more weeks until Emily accused me of lying about the truth that I told her. So I asked her if she asked the other night shift employees to verify or deny my story. She said she hadn't. I had absolutely nothing to hide. I told the complete truth because it was the right thing to do.

I quit again and emailed her a list of all of the people with phone numbers she could ask to verify my story, including the phone numbers of my family. Not only did it include what I did wrong, but it included the pattern of disrespect I was shown throughout the season. I emailed it to HR, both of my bosses, and the superintendent.

I was not the only person disrespected in multiple ways at this job. The upper management got away with: sending people in UTVs when there was lightning, sending people in vehicles that could break down in the backcountry, refusing to rescue patrollers when their UTVs were being overtaken by high tides, and much more. I witnessed Ashley yell at the top of her lungs at my co-workers for the mail not arriving and even at Emily for having a hair appointment. The hero title I once associated with Ashley had turned to *bully*, a sentiment echoed by several of her past employees.

I had to message my co-workers and explain my side of the story, something which I found quite embarrassing. Upper management had a reputation for spreading rumours and I wanted to be ahead of them. I met a lot of wonderful people that I highly respect and I thought that confessing would be the best way of showing it. Luckily I was met with a lot of support from my co-workers, who agreed that I had taken the brunt of abuse from Ashley and Emily.

I wrote my resignation letter so that no other biotech would be crushed the way I was ever again. Did I break the rules? Technically yes, and no one will ever punish me the way I still do, over a year and a half later. But I didn't deserve to be treated the way I was, which has been a trend on their end. I've been told I shouldn't feel haunted by my decision, but I do. I cringe every time I have to click *"you may contact my previous employers"* on an application. How did my dream job with my hero end up being the reason I had to go to therapy?

But my heart was in the right place. I made sure everything was as safe as possible and finally did something at that stupid job for myself. I've spent my entire career and most of my personal life doing things for other people or animals. That day, I finally made sure I got what I knew I deserved to see. I wanted to see my babies, the ones I checked on every night, I helped find, transport, and account for the start of their new lives. I wanted to see that all of my janitorial work, looking past sexual harassment, and the disrespect I went through, had helped in the conservation of those little hatchlings.

*"Do you think your bosses were being a bit sexist?"* My therapist asked.

*"No, they were both women."*

*"Maybe they felt threatened by you?"*

*"I mean, they are both really established in sea turtle science."*

*"Do you think you were taken advantage of?"*

*"Yes. But also, I feel like I was forgotten."*

*"And then when they had to blame someone for a faulty fire alarm management did not fix, you were finally remembered."*

In no way do I aim to disrespect the women I worked with. Without them, this sea turtle population would be almost gone. I thank you for giving me the opportunity and respect you for your work. But I respect myself more than the way I was treated.

# | 11 |

# Finding home

## ABIGAIL SMYTH

The idea of home has always been tough for me. Growing up in South Sudan, where my parents worked rehabilitating refugees, I was very much a 'free-range' kid. Running barefoot across murram roads, splashing in monsoon puddles and sculpting creatures out of clay dug from the earth, I enjoyed the world around me.

So you can imagine that moving to Ireland at the age of six was a massive cultural shock. Outwardly, this is where I *fit in* best. My parents are both from the UK and I am very much white. People assume I belong here. But as a kid, I really struggled to assimilate, refusing to wear shoes or play with the toys that other children loved. I remember once being asked to draw a house as homework and, after submitting a picture of a traditional Sudanese tukul, being told that I had done it wrong and to copy a friend.

I found it interesting how, in Ireland, the culture is to view nature as a recreation, going on hikes and picnics, while in South Sudan, most people have a direct reliance on nature, collecting food, water and firewood as a means of survival. My teacher had invalidated a mud house with a grass roof, supporting the idea of a more industrialised and less natural building. In Western cultures, we're very quick to label people as poor or disadvantaged because their lifestyle doesn't match our economic values while we live vastly unsustainable and destructive lives.

Growing older, and settling into life in Ireland more, these thoughts weighed on me more and more. Sometimes I found it hard to relate to my friends and their childhood experience. I found comfort in the fact that my true home was thousands of miles away, in South Sudan.

Soon, my family decided it was time to revisit where we'd once lived. I was in my teens at the time and extremely excited to finally return home. The sad thing is, we can't chase the past. And as much as I'd relied on belonging in South Sudan, when we arrived there, what awaited wasn't home. The idea was that people would understand me, relate to my life experience and that nothing would have changed. But the houses I remembered had been burnt to ashes, the mud I'd once played in was replaced with concrete and I stuck out like a sore thumb. There were even language barriers now, words I could no longer understand.

I was lost.

Then one night, I was sitting outside on the veranda feeling incredibly lonely when I heard the calls of fruit bats. They hunted gracefully in the dark as I watched. It was just such an awesome moment, and I realised that home was always going to be in nature, not a dot on a map. It was an incredible feeling of belonging to a wider pattern of life and I finally had my place.

Now, I'm almost 20 and about to begin a marine biology degree, ready to advocate for conservation, for my home and the homes of everyone at risk of becoming a climate change refugee.

# | 12 |

# Lorises and roller-coasters

## SAPPHIRE HAMPSHIRE

*Java, Indonesia. October 2017.*

Picture this: You're on a mountain, a steep mountain. It is 4 am, and you wake to the sound of the call to prayer from the mosques echoing down the mountain.

You drift back into a light sleep waking up again later to the revving of scooters as they whiz past the field house towards the farmland.

You hear the caged songbirds tweeting, then the chitter-chatter starts, and then, you hear more revving of scooters.

You walk downstairs and greet the cleaning lady and cook with a cheery *"Selamat Pagi Ibu!". Good morning, ma'am.*

As you eat your breakfast you smell rice cooking and hear tofu sizzling in the pan of palm oil.

You take off your socks and head into the wet-room. Water is running in the water bath as you brush your teeth, you step across the wet floor, you hear again the revving outside, people shouting on a megaphone. *"Ah, it must be Friday,"* you think; they're playing sports in the village.

You hear a rumble of footsteps and giggles as you enter the lounge. *"Miss! Miss!'* they shout as several kids pop their heads through the door. You walk out and colour with them. *"Miss! MISS!!".* You tell them in Indonesian you have to go, but hand them paper and pencils and set them a colouring challenge.

You walk down the hill and jump onto a scooter and head down the mountain. *"Buleh! buleh!"* people shout as they see you, kids run up to you in the shop *"Buleh"*, *"Hello!"*; you pose for several photos.

On your way back up the mountain, you feel the breeze hit your face as you hold on for dear life during the steep gradients of the hill. Back home you hear the revving, birds tweeting, call to prayer, chitter-chatter, revving again, chitter-chatter, dogs howling, revving, *"Miss, miss!"*, the call to prayer, *"Selamat Siang!"*, shouting, revving, call to prayer.

Suddenly it is 5 pm. You head to bed and sleep until 9 pm, then eat some dinner. You now hear a distant revving; the chitter-chatter has died down and some dogs are still howling. As you get dressed, you hear the rain start to pour, so you shove a spare coat into your bag. You grab your water, your head torch and you venture out into the rain.

Underfoot you feel squelching in the mud, you step over rocks, and you feel your thighs strain as you walk up the mountain. You meet the first shift team, track the loris and start your shift.

Squelching, rain, *"Can you see them?"* clambering through the rocks, climbing through thick bamboo, *"Beep beep"* from the tracker. *"Buzzzzzz"* those mosquitos are coming, track through the mud some more, clamber up a cliff face. More rain, you hide in a shelter, *"Ah shit, my stuff's wet"*. Switch coat, keep tracking, *"beep beep"*.

It's suddenly 3 am.

A cool breeze sweeps through and the tracker starts a fire, *"crackle crackle"*. It is all calm. The loris is in its sleep spot for the night. You look around, you see the stars, you try to stay warm and at that moment, you feel calm. A sense of peace, as you wait for the inevitable call to prayer to start again at 4 am. You walk back down the mountain, the sun is starting the rise. You hear the cockerels call as you head into the house. Covered in mud and sweat, you quickly rinse your arms and hands and peel off your layers.

Snuggling into bed, you shut your eyes just as the world begins to come alive again.

***

I'm Saphy. I'm British but I focus on conservation globally. It may have seemed a bit odd but I promise my introduction serves a purpose.

This was a typical day during my internship in 2017-2018 where I spent eight months on the island of Java, Indonesia studying the critically endangered Javan slow loris. So why am I telling you my story now, and in this way? I have spent a large amount of my life struggling with General Anxiety Disorder, mild OCD and hypersensitivity to sound. Just before travelling to Java, I had been diagnosed with an additional trauma-related condition.

It has taken me years to understand why I spent a lot of the time on edge over there. Despite the same routines in the village occurring every day, nothing was ever truly the same. It also felt like nothing ever stopped and everything was constant. But considering my mental health, the constant shifts in the environment, those slight changes, the amount of noise and external stimulus, I was processing on a minute-by-minute basis that clearly kept me anxious, but I didn't really realise that at the time.

The placement gave me some amazing skills: I got to see slow lorises in the wild and help with conservation. I got to meet incredible people, teach at schools, and engage in public outreach. Truly awesome. But for six months after my placement, my view was very different from what it is today.

I was desperate to leave, I spent days crying, I felt anxious 24/7, I always felt like I didn't know what I was doing and when I did take charge or initiative, I would knock myself back down (An irrational fear really: my supervisor was a kind and amazing woman and I have a lot to thank her for). I couldn't drown out the noise, and in a house full, sometimes too full, of people, I often felt left out, or isolated and alone. Leaving Java I was bombarded daily with many *"How was it? It looked amazing!"*. Yet I didn't want to talk about it; all I could remember were those moments where I couldn't escape, where everything was loud and I had felt trapped.

The culture shift was hard to process after eight months away. I was back in a world where people didn't look at me or talk to me, where scooters and bikes were not the common vehicles and I hadn't heard a mosque in weeks. That also felt strange. I felt like I'd been picked up out of this world I lived in and that maybe, the whole experience had never even happened. I used to dream I was still there and wake up dreadfully confused. In the months after, I still had to finish my project and worked super hard with the data that we had collected from the camera traps. In the end, I was proud of my work and I still very much am. Finishing it also allowed me to feel a sense of completion, in which I started to notice my positive memories coming back.

I could now list a million and one things about why my placement was so valuable, both as a budding scientist and for my personal growth. I could also list many amazing memories and positives about living where I did. But I want to tell you how hard it was to accept all of the positives because my own mind felt entrapped in prison.

I felt like a lonely conservationist living there, despite being surrounded by people and never alone. I also experienced my estranged father dying and my grandma back home passing away and being unable to leave and be with my family, it took its toll on me. I felt so far, *so* far away from home. I have never been a homesick person, yet I made my mum cry on Christmas Day because I told her all I wanted was to be at home.

Looking back, I have so many fantastic memories surrounding that moment and that day was an exception for me. I didn't hate it, but by the end, it had just made me upset which is why I had called my mum and my brain had fixated on that. It took me a long time to stop associating all my Instagram photos with the negative memories, but actually, the positive memories associated with why I took the photos.

By my fourth month in Java, the furthest from the village I'd been was the local town which had a pizza restaurant with free Wi-Fi. One day everything came crashing down around me.

I had fallen during a shift and hurt my ankle. I had been stuck at home unable to do anything for days, and one day, I lost it when I found one of the house cats had pooed in my bed. I sat crying for hours. My supervisor sat with me, talked, listened, showed how much she cared and her kindness. A common theme if I think about it. My supervisor not only worked insane hours but made the time to laugh, joke and manage/supervise me – as well as others. She had been there through both family deaths. I connected to her, but with that came the anxiety of letting her down. She wasn't just a supervisor to me: she was my family and I never wanted to disappoint her. There were times of friction due to high stress levels from both of us. But in the end, I can't imagine a world without her and what she has done for me.

The feelings of being left out dissipated when, over Christmas, our team got smaller and I connected with my fellow intern. You see, a lot of these feelings of anxiety, stress, not enjoying myself, made me feel ungrateful. I recognise how lucky I was to have the opportunity to go to Java and work on such an incredible project, and I realise how great it was that I could get a high enough student loan to afford to live there and be there. Therefore, when I felt crap about it, it felt like I was being ungrateful and that, in itself, made me feel so much worse.

This was the same dilemma that I experienced when I was home, because I couldn't, for a while, remember the positives, but could not tell people the negatives without the risk of sounding ungrateful. I can now happily acknowledge both and am confident enough to explain *why* I struggled and not put the blame on other people, the project, and the place because in reality, that's their life. I did fit in and there are aspects I truly miss. I enjoyed that although things were constantly happening, it was actually pretty relaxed. I miss the people and the kindness that they showed me when invited to large events or when I lived briefly with my Indonesian teacher and his family. I miss speaking another language and sharing stories and riding on the scooters down the mountain in a poncho because it was raining so hard.

But on those days where I did feel anxious, all I remember are the noises and the feeling of isolation. I would fixate on them. I would find every step torturous if I was walking through mud. I would find every call to prayer so loud that I couldn't concentrate. But there were so many other days where none of that mattered and I would be out smiling laughing, playing with the kids, out on day shifts saying hello to the farmers. I spent a lot of time looking at camera trap photos and I enjoyed it.

***

I find it incredible how our brains work when we have trauma or anxiety or hypersensitivity – because I can read my introduction in two very distinct ways. The first reading is with a positive outlook on a day. I can think about the creative way I made my hoodie wrap around my face to avoid mosquitoes, how I enjoyed the smell of the cooking during the morning, how waking up to the call to prayer at 4 am meant I got to learn about the locals' relationship with their religion. I enjoyed the sounds of the scooters as I whizzed down the mountains and getting to speak with random people and kids in Indonesian.

On the good days, these sounds and actions were good and I enjoyed the day.

On anxious days, however, my introduction sounds like a nightmare to me. The constant stimulus, environment change, the sounds, back-and-forth, sounds, more sounds. Overwhelming. No wonder I struggled on those days. But I could never really pinpoint it and thus, could never understand how to help myself. I didn't practise mindfulness because I hadn't worked out how. But I had unknowingly taken some of those important steps, without realising.

For example, the language. Learning Indonesian really helped. I think I'd have been lost without it. Being able to talk to the people who drove me on the scooter down the mountain, or the kids in the schools made me feel I could connect. The more I learnt Indonesian, the less trapped and lonely I began to feel. I also took a break and went on holiday to Singapore and Sumatra and saw my mum and that change and that slight feeling of home helped push me through the last part of the internship.

Living in a small, Muslim, conservative, and traditional Sundanese village means you have to change your way of living to respect the culture. Although I was comfortable doing so, it could be challenging sometimes.

I am the first person to tell you that I grew up privileged. My family weren't comfortable with money during my childhood, but our background gave us more systematic rights than several other people deal with. My mother was incredible and did everything she could for me regardless of money.

Therefore, I grew up in a fairly wealthy area, though not wealthy ourselves, with a bursary to a private school, and for many years, I ignorantly reaped the benefits of privileged white girl life. Travelling from the age of 13 to countries further than Europe helped open my eyes at an early age. Therefore, my ability to adapt to different lifestyles is pretty good, and I quite enjoy experiencing other cultures, because I think it is super important, but this doesn't mean that sometimes I just want those home comforts back. I didn't know how to cope with it at the time.

I think I have learnt a lot now. That was my first time away without seeing my family for more than six months and it shook my world. Despite being an adult and 22 years old, I felt like a baby, but I realise that is important to go through. As I had to grow up into my previous life with my family, when I moved and changed cultures, I had to grow up again into that life. It was hard, but it was worth every second; I would never regret any of it. I don't see the point in regrets anyhow and I think Java did me a world of good and I felt immense relief when I began to go through that realisation.

I was then able to share my stories and memories, and I actively started to learn from the experience too, sharing both the positive and the negative.

I've always enjoyed my alone time, but in a privileged manner, so to speak. It's why I always preferred the "second shift" – one where I'd go out from 11 pm until 5 am because on these shifts, I felt most at peace, even if chasing a loris or hiding in a shelter from torrential rain. The actual shifts were my focus. It was hard to climb the mountain and navigate the fields and chase lorises for hours – but it's what I loved doing. I loved it. I loved seeing them, recording their behaviour and finding happiness. I enjoyed the quieter nights where I could see the stars and the only sounds were from me, one of the trackers and the devices we used. At other times I also found peace in being alone, for example, I spent New Years' Eve on the roof watching the fireworks down the mountain and I called my family.

However, there is a major difference between feeling lonely and having alone time.

The feeling of loneliness was draining, negative and exacerbated all my self-doubts, worries and insecurities. For months, the idea of being alone drowned out the reality of actually having a tight-knit connection with the people there. When I did open up to my friends at the project, I realised that everyone has these feelings. It's normal. It's okay to feel it, but we have to find a way to cope and sharing how I felt really changed things for me.

I also had the pleasure of meeting some incredible people. If I'd have succumbed to the earlier feelings of wanting to leave, I'd have never met these amazing people – and they changed my life for the better. It's just so easy to get wrapped up in our heads, particularly when we spend our days in perpetual anxiety, triggered at the small things and therefore unable to cope with the big things. I choose to acknowledge that I felt both amazing and shit during my placement. Because in reality, that's the truth. Sometimes it felt like hell and I wanted to leave and I was crying, but other days, I laughed my head off and can't imagine having not been there. Mental health is stigmatized still: I still find it hard with my current endeavours, but going through my experience in Java has led me to grow into the conservationist I am now.

The world of conservation can feel brutal and helpless at times. That is an important lesson I learnt whilst over there. It was hard on some days to remain positive.

There was one night I distinctly remember watching one of the lorises we tracked walk on the ground for 11 minutes (they are arboreal mammals) and I just cried. Knowing that human presence has destroyed the habitat is heart-breaking, but I had to realise their livelihood mattered, too. Thus, the next day, I likely remained pessimistic.

But there were other days that sparked joy, like when I found a loris caring for a new baby or when I interacted with local children in the village who have started to understand why it is important to save them. I think working out in the field for long periods, especially in remote areas, is difficult and I think it is a rollercoaster of emotions, particularly when coupled with a lot of sensitivity with mental health problems. However, without Java, I doubt I'd be half the person, half the scientist and half the conservationist I am today.

Whilst I have learnt a lot, grown a lot and learnt how to cope with many of the issues I faced back then, I still struggle. I doubt the struggle will ever fully go away, and I'm okay with that. Knowing I was strong enough to cope then, has led to me coping through things I never knew I could now. I have chosen to speak out about anxiety now, and I try my best now to share my stories to help others not feel so alone, often how I've felt during my life.

Being a conservationist takes guts. We are setting ourselves up for a higher level rollercoaster than we could choose for a life, but I believe it is worth it. Because we are fighting for the environment, for the animals which can't fight for themselves and despite feeling like a lonely conservationist, I know that we are united in how we feel. I shouldn't have to walk away from an experience and only say the positives and hide the negatives so I don't feel shame. I think we should be embracing that conservation can be bloody hard, as long as we embrace the good too – even if sometimes that takes a while and a bit of work from our brains to accept.

I hope that you could follow my rambling. I have to point out that my story is focused on how I coped with my mental health in the field, not the work I did there. However, I feel it's important for me to just give you a heads-up if you choose to look up what a slow loris is. You can search the animal on the internet, but be sure to avoid any cute, pet-like videos – these are often illegally traded pets, and the animals are usually stressed. Instead, I'd recommend checking out the main conservation work happening.

*** 

It was difficult to revisit some of those emotions when writing my story. Obviously, this snippet of my time there does not cover everything: there are a million more amazing moments and additional struggles I faced. But some things I am choosing to keep to myself, for my own sake. I want to reiterate that this experience is the reason I am the conservationist I am today, despite starting that journey many years before. Therefore, although I have several conservation-related stories, I felt this one really needed to be told. And in the end, it has set me on a path I want to be on which is conserving tropical nocturnal mammals. That's my goal, my dream.

With my mental health, I know it can be difficult to cope at times, but it isn't about me, it's about the animals I'm trying to save. That is what led me to survive my challenges in Java and what pushes me to this day. I am truly grateful for the project, for the people I met and the experiences I had out in Indonesia, and I look forward to going back one day as I really cherish what it has done for me.

I hope you all can find strength while facing hardships on your paths and journeys and hopefully, by confiding in one another, we can choose to be not-so-lonely-but-still-sometimes-lonely-conservationists.

# | 13 |

# The curious case of "career loneliness"

## TIRTH VAISHNAV

The idea of being alone is not inherently scary for me. The idea of *feeling* alone, however, is an entirely different beast.

Being the youngest in a joint family of seven and having grown up in Mumbai - one of the most crowded cities in the world - to get some alone time seemed like a blessing. I have always cherished solitude; being by myself gives me a chance to think, have an internal dialogue, ponder the deeper meaning of existence (no, seriously!), or just be still. It replenishes my energy for another bout of human interactions. Solitude comes naturally to me so I don't often feel lonely in day-to-day life, regardless of whether I'm living alone in another country or at home with family. The kind of loneliness that I have experienced, however, is what I'm calling *career loneliness*.

My family is the artistic kind. Mom and dad encouraged us to take up various hobbies in our childhood. My siblings were extremely boisterous: they enjoyed dance classes, karate, drawing, sports. I was the shy one who did whatever they did, and so, I didn't stick with most of those hobbies (except dancing, through which I found my artistic expression). I was always more scientifically inclined but had no one to share my interests with growing up. As a result, I kept my love for the natural world to myself. I could have suggested trips to nature parks or safaris. But to fit in, I continued with the borrowed hobbies and got my fill of nature only from National Geographic and The Discovery Channel on TV.

Come to think of it, my career loneliness began much earlier than I realise. In the early 2000s, I was living in a concrete bubble, having no exposure to the natural world (these were times of dial-up internet, so it wasn't easy to make 'connections', pun intended). Most budding naturalists in high school would be enjoying backyard birding, nature trails, volunteer positions or amateur photography. Instead, I would resort to reading books about animals and nature and keeping checklists of various species off the internet. But I was struggling with trying to articulate where I wanted to go from there. The closest I got was in the 12th grade when I proudly declared to my family that I want to study *something related to animals*.

This came as somewhat of a surprise for my parents. Me, who was creeped out by the gecko on our kitchen wall; me, who had never expressed any interest in outdoor activities; me, wanting to study animals, nature, the great outdoors! Naturally, they had to figure out if I was being serious, so they took us on a trip where we did all the things that I love. We went to the Madras Crocodile Park where I clicked pictures holding baby crocs and pythons, and for a safari at the national park. My enthusiasm convinced them that I knew what I was talking about. But my seclusion from the world of wildlife and nature lovers was such that I did not know what the next step was. I decided that I would pursue veterinary medicine, which seemed like the most obvious choice at the time. Having no local connections, and a desire to *break free* and *become my own person*, I got admission for BSc in Animal Biology at the University of Guelph, Canada in 2009.

At Guelph, I was introduced to the wonderful world of research. I realized that I was drawn to academia, and I finally found a name for what I wanted to be called professionally, so I switched my major to wildlife biology. It was all well and good until I graduated in 2013 and started looking for the next thing. Since I was fairly new to research, I didn't feel confident enough to start postgraduate studies immediately. My visa prohibited me from applying for most biology research positions that were restricted to citizens and permanent residents. I ended up doing (unpaid) internships in wildlife rehabilitation and reptile care. These positions did contribute to conservation (local wildlife preservation and conservation education, respectively) but took me far away from my research roots.

I didn't have a professional guide to help me sort out these feelings, or friends who were inclined towards academia rather than zoology careers. All I wanted was to sit down with someone for a coffee and untangle my aspirations for the future. And of course, there were the unsolicited remarks by relatives asking if I'm ever going to stop studying and get a job, or just plainly advising me to change my field since there are no financial prospects in academia. Anyone in an unconventional profession (especially Indians) would be able to relate to this.

This was probably the time when I felt most lonely. I was living alone in a foreign country, broke, directionless, and dejected. It took a toll on my self-confidence and sowed the seeds of the dreaded imposter syndrome. I decided to reboot my life, so in 2015, I returned home to India to fix the physical loneliness problem which would help me deal with my career loneliness. By this time, I felt like an outsider in my own country. The academic scene had progressed considerably in my absence: people were more connected and new postgraduate degrees were being introduced for wildlife studies even in big cities that were previously impenetrable for such majors. I felt like a fool to think that I had any kind of advantage just because I had an undergrad degree from abroad because I had no prior research experience.

Cue the intensification of imposter syndrome. I didn't know where I belonged.

There were extenuating factors that resulted in a gap of four years between my BSc and MSc that I ended up pursuing in Mumbai itself. The curriculum was course-based but I was able to dig into my research roots and share that knowledge with my peers. I also learnt that there is a huge difference in the mindset in what *wildlife studies* mean in India from what I was used to in my undergrad. In Canada, my training was heavily focused on research methodology and critical thinking. But because of the wildlife situation in India, with its high biodiversity and human-wildlife proximity, the approach was much more hands-on, action-oriented, and inter-disciplinary. There wasn't as much emphasis on scientific writing and experimental methods.

My MSc dissertation research took place in 2019 in the most rural areas of central India, in the fabled forests from *The Jungle Book*. It was my first true independent research experience and the first time that I felt like a real conservationist. I even got my first publication out of it. There was a feeling of fulfilment like I had never experienced before. Upon graduation, I started applying for PhD positions. It took many months and every ounce of patience to finally find the right fit, but then the pandemic hit.

Along with the rest of the world, I went into survival mode and put everything else on the backburner. When I got an acceptance email from Victoria University of Wellington in New Zealand, I didn't know how to celebrate. I wasn't sure if I would be able to follow through with the dream, but thankfully, the world of online and offshore education opened up for students across the globe. I was able to start my PhD remotely from home. My supervisor and support staff at the university were exactly the professional mentors I was looking for in my entire student life.

But even now, the career loneliness still rears its head from time to time. There is a sense of insecurity in seeing your classmates from university advance in their corporate jobs, buy houses and settle down. Also, having limited experience with independent field research, the idea of starting a PhD sitting in my room in another country seems ridiculously unfair (and daunting).

Knowing that there is a teaching assistant position and an office desk with my name on it, waiting for the borders to open.

Not being able to walk into my colleagues' office to discuss ongoing projects and research gaps.

Not being a part of the wider campus community, feeling like an imposter playing at ecology from the confines of home.

I often feel insecure about not having done enough in my field as a researcher, but I don't know what the yardstick is (*is there even a yardstick?*) Isn't the fact that I'm doing a PhD in ecology proof that I have somehow paid my dues and navigated through this labyrinth? Is financial prowess the only barometer to measure success? But in feeling these things, at last... *I know that I am not alone.* Countless students around the world are sailing in the same boat. The pandemic has been a giant reminder to count our blessings and wait for things to fall into place. Isolation has taught us that loneliness is not natural to humans, who are social creatures.

It is important to reach out, to take the first step and seek the connections that we desire. Even though my career has not been as straightforward as others in the field, I am on my way. The connections I have been longing for are finally starting to form. The pandemic will end, nature will heal as it always does. As I enter my 30th year, my academic journey is just beginning. The road may have been longer and more winding than that of my peers, but that's ok. I am no longer the little kid with a dream and no one to share it with. I hope that no budding conservationist ever feels the '*career loneliness*' that so many of us experience at various stages. As the world grows smaller, our community grows bigger!

# | 14 |

# My life in the hands of birds

## SEBASTIAN MORENO

You'd think having a gun flashed at you while doing fieldwork would get you to pack it up, go home, lock yourself in your room, and reconsider your career choices. I won't pretend to be brave and I will admit: I did finish fieldwork that day, go home, and have a good long cry. I thought nothing could go wrong with becoming an urban ecologist since it was a perfect blend of nature and city.

I was born in New York City and I wasn't exposed to nature growing up. When I was twelve years old, my family moved to northeast

Pennsylvania. I had never seen that many trees before! It was beautiful! Our backyard led into the woods and my parents were always imagining my demise if I went outside. As an adult, I still get the occasional:

*"Cuidado, qué se lo come un oso! / be careful that a bear might eat you!"*.

I started hiking in my teenage years and didn't really get involved in ecology until halfway through my third year of college.

After completing my undergraduate degree, I felt like I lacked experience and knowledge, so I went back to school to pursue a master's degree. Knowing I liked working with birds, my advisers and I came up with a project looking at a vacant lot in St. Louis, Missouri and how they impact wildlife. Before Missouri, the furthest west I had ever been was Pittsburgh, Pennsylvania and I knew nothing about St. Louis and Missouri except for what I heard on the news.

For those not familiar with this city, let me give you a quick history lesson. The city of St. Louis succeeded from St. Louis county in the late 1800s, creating fixed boundaries. Fast forward a couple of decades, and we have a fluctuation of socio-demographics due to racism and segregation. In 2014, Michael Brown was shot and killed in a town right outside of St. Louis that only further split an already torn city.

My research took place in the northern city, in two neighbourhoods with the highest concentration of vacant lots. I was tasked to identify vegetation structures on these lots and conduct bird counts. Once or twice, I did have some company but, for the most part, I did everything alone. Being a person of colour and from the city, I thought this would be easy.

Day one was a bit of a cultural shock for me. I had never seen, first hand, the injustices occurring. I was in the middle of an area of high crime and neglect. During my time in the city, I stumbled upon lots of birds, as well as dealers ranging from sex to weapons and everything in between, and plenty of crime scenes. Though the rest of the city had forgotten about them and labelled their home a war zone, there were still plenty of residents who hoped for the best and wanted change.

Summer of 2018. I was doing my bird counts. Just a regular morning walking around the neighbourhoods with my binoculars and clipboard, scribbling any bird I could identify. After two years of walking around the streets, I had developed a small reputation as *Bird Man*. Most people knew that I wasn't a threat if they saw me walking the streets. A kid comes out of his house and walks down the middle of the street with a roll of money in his hand. I don't know where he is going, and I didn't want to find out. Unfortunately, the path he was taking was the same as my transect. To add to the series of unfortunate events, it looked like I had been following him for the last 50 meters. One of us had noticed this, the other was too busy staring at birds.

The kid turns around and lifts his t-shirt and, on his waistband, sits a gun. My mind goes blank. I just remember his high school's *Class of 2019* t-shirt and thinking:

*"I'm going to get shot by a 17-year-old who thinks I want to steal his money. When all I am trying to do is find out how many freaking birds are on this block!"*

My mom always warned me that being outside would lead to my end. Had she been right?

That moment I perfected my elevator speech about my project. Rattled it off in one giant breathe and hoped this was good enough. It was. He let me walk away and I was going to be able to bird another day.

I'd be a liar if I said I didn't grow slightly jaded after that. I was upset I was conducting scientific research in an area where science was pretty low on their priority list. But the show must go on. I needed to collect data so I can get my degree.

One morning, while doing my counts, I came across a kid waiting for his bus outside his apartment building. Don't worry, there was no gun involved here. He was about eight years old. He remembered me from the last time I was around and said he had been waiting for me. He was hoping I would come around again so he could show me the different birds he had spotted around his complex and the various nests he had found.

Through all his excitement, I only thought it was fair that I asked him to join me to conduct my count that was in his complex. After our bird count, we had a pleasant nerd off about birds and I walked him back to the bus stop.

That small moment had a big influence on what I want to do. I realised how important it is to reach underserved communities and provide them equal opportunities and access to nature. I know what sparked my interest in ecology, I'm sure all of you can also recount your tales. I want to be able to provide these moments to young individuals. This fall, I will be starting my PhD and will tailor my dissertation to work with underserved community members. These kids are our future and we need them to continue our work, and frankly, our fight to save the planet. I would be proud and happy if I can influence just one child to pursue a career in science and not follow the same path that others in their neighbourhood have taken.

# | 15 |

# For the shihuahuaco

## ELENA CHABOTEAUX

In a remote area of this planet, deep in the Amazon rainforest, a majestic 60-meter-tall tree is struggling to survive and it's about to face its own extinction. Due to deforestation and illegal logging, many important tree species are disappearing, leaving parts of their giant trunks and their stumps all around the "*Selva*" (forest).

My name is Elena Chaboteaux and I'm a 24-year-old conservationist from Italy. My current research involves the conservation of a giant sentinel of the Amazon rainforest: the Iron Wood tree (*Dipteryx micrantha*), or Shihuahuaco, as locals call it. This species is rapidly disappearing, being cut down at high rates due to intense illegal logging activities; one individual takes 300 years to reach a diameter of 50 cm and almost 1500 years to reach a diameter of 1.70 cm. The Shihuahuaco lives for *hundreds* of years and during its lifetime, one individual could witness both the Crusades and the first human on Mars. This species has undergone a number of cuts, exceeding 270,000 in 10 years, an average of 74 Shihuahuacos cut down per day. It is believed that the species could disappear from at least two regions of Peru by 2025 and in less than 10 years, it could go extinct.

Here I wanted to share with you the story of how I fell in love with this imposing yet very vulnerable tree and how it has been able to change my life, not only as a conservationist but as a person.

Six years ago, I was watching a documentary regarding illegal deforestation in Brazil and Peru and I clearly remember how powerful it was. There was a particular scene in which we witness the growth of a Shihuahuaco seedling and, as time passes, you see that the same individual was there: when the Vikings reached France, when Marco Polo travelled through Asia along the Silk Road, when Columbus set foot on the American land, when Copernicus first proposed a heliocentric system, during World War I and II, during Mahatma Gandhi's non-violent independence movement, and was still there when Neil Armstrong stepped onto the lunar surface.

It was there until 2012 when a couple of people saw it and cut it down in less than 45 minutes, to sell it as a high-quality coffee table in the U.S, China and Mexico.

More than 1,100 years of history, questions and answers. Completely erased in less than an hour. I already knew a few things about the issue, but that scene had me in tears. I spent time educating myself more and more until 2018, the year in which I had finally managed to save enough money to travel to Peru. I decided to intern for the Madre de Dios region of Peru to study wildlife and ecology, some *Pteridophytes* genera and take notes and pictures of some plant pathogens that I could study for my MSc thesis. One of the main goals was indeed to find Iron Woods.

I never would have thought that my first encounter with a Shihuahuaco would occur on my way to the field site on the Las Piedras River, the day after I landed in Puerto Maldonado. I was in a car with three other volunteers and naturalists, following a dirt road in the forest when a massive truck, coming from the opposite direction, made us stop and pull the car over to let it pass. That truck was carrying huge tree logs. The researchers working at the site told me they run into dozens of trucks like that one, every day, carrying the trunks of logged Shihuahuacos. That's when I realised that everything I had read was true.

Luckily, a few days after that first shocking encounter, I saw my first standing individual. A huge, 1000-year-old tree that made me feel as if I was an astronaut, floating in space and watching something that the whole world should watch. Exploring the vast unknown and feeling like a tiny and scared dot compared to thousands of years of history and natural events that one can't even imagine.

I held my breath and told myself: *here we are, Elena.* I knew I would remember that moment for the rest of my life.

In 2019, I went back to Peru to carry out my MSc thesis project: the conservation of *Dipteryx micrantha* through some molecular analyses on the bacteria and archaea communities living in its rhizosphere. To have a broader view and compare samples, I needed to find both standing and cut down individuals. Thanks to some GPS coordinates kindly offered to me for the purpose of this research, finding the standing ones wasn't too hard. Once you are close enough to the point, just look around and you will easily find the tree. Finding the stumps of the logged ones was way more difficult because I didn't have points on the GPS, just thousands of hectares to explore.

After days spent wandering around following old logging roads with no results, two forest rangers on patrol showed me the location of the first stump which was pretty much on the trail they used to walk every day. Because the whole thing was a favour they did for me and therefore we were all in a bit of a rush, there was no time for feelings. Maybe because I wasn't alone, the sight of my first *dead* Shihuahuaco wasn't too shocking. Five days later I found another old logging road, almost completely recolonized by vegetation that would have been impossible to walk without having a machete. I followed it for a while and then went *off-trail* for more or less 15 meters from where I was standing, I saw something. The closer I got, the thicker the vegetation became.

And there it was: a *huge* Iron Wood stump, covered in leaves, plants and shrubs, totally forgotten. Small trees and bushes all around it had already replaced the empty spot in the forest canopy, acting as a natural roof, sheltering the stump from direct contact with rain and sunlight. It truly felt like I was standing in a place no longer existing, a place that used to be magnificent: I was an uninvited guest attending a never-held funeral. I got there by chance but I wanted other people to be aware of the existence of this tree and the way it made me feel, so I sat there and wrote a bit about it, with tears in my eyes. Then I left.

The whole experience made me realize that being so passionate is exactly what makes us upset, frustrated, depressed and anxious, but it is also what makes us so lucky. The more I care, the more I struggle with anxiety; only recently I started to understand it is because I am so in love with my job and I want to be great at it. I spent 2019 fighting 24/7 to go back to Peru with my research project. It was the most difficult and stressful year of my life but I didn't give up mostly thanks to a couple of trees I saw in 2018. Isn't that weird?

The best thing about passion is that it takes so very little of what you love to charge you up for so long. Even picking up a Shihuahuaco seed, handling it and thinking that this seed, that fits on the palm of my hand, could become a thousand-year-old, massive, 70-meter-individual. That is something that gives me goosebumps and motivates me so much.

Those Shihuahuaco trees and so many other things that I have seen and experienced in the rainforest are what I'm fighting to preserve right now, and what gives me anxiety, depression and sometimes, the feeling that I am not good enough. I'm sure the path ahead of us will be challenging for passionate people, but I believe that is also what makes life so beautiful and worth living.

# | 16 |

## Tribute to max

### JENNIFER L. HARTMAN

I am approaching my 15-year mark working in the field of conservation biology, surveying sensitive species around the globe. The work has been gruelling and gritty and, for those of us in this field, it is often a solitary toil.

But for me, I have never felt particularly lonely or alone working in the field. Until recently. Conducting fieldwork suited my introvertish love of the natural world. I have to admit though, I was not completely alone. There was Max, my 'co-woofer' and conservation detection dog. These smart, energetic canines are much more than essential "equipment" in this work - they are teachers, companions, confidantes, and our best friends.

This story is dedicated to those who work with animals. To the dog handlers, mushers, zookeepers, horse whisperers, and to anyone who understands the intimate connections we establish with animals.

I have been fortunate to have conducted research in Nepal, Cambodia, Mozambique, South Africa, and British Columbia but also along polluted streams in New York, or bushwhacking through overgrown and rugged Oregon mountains, and surveying lonely highways in Utah and Wyoming. This career path typically requires a degree in biology and extensive fieldwork experience. However, my degree is in English Literature. I had lofty plans to pursue graduate school in wildlife biology so that I could continue to climb the proverbial ladder, but then I met Max.

Max was a spunky, white-whiskered, paper-eating, sweet-natured rascal. He was also obsessed with playing fetch, a requirement for all of the Rogue Detection Teams (RDT) dogs. RDT is a conservation detection dog program currently with 16 fetch-obsessed, high-energy detection dogs, all rescued from shelters or as owner releases. We work with the dogs to be scent detectors of scat, toxins, plants, or other animals, and then deploy them on projects with their bounders (handlers) to locate specific data. My dogs and I have surveyed for species as diverse as African lion and cheetah, pangolin, storm petrel, wolf and cougar, and even orca (yes, orca scat floats, if just for a little while). I've lived in bush camps with chatty hyenas as neighbours, backpacked through the stunning backcountry of places like Yosemite National Park, slept in jungle hammocks in Cambodia, battled blood-sucking leeches in Vietnam, and taken all means of travel, from helicopters, snowmobiles, and rickety boats to arrive at remote field destinations to survey for cryptic odours.

When I first met Max, I was a new trainee on what was supposed to be a temporary four-month summer research study to survey for northern spotted owls in California using detection dogs. The goal of the work was to learn about the habitat use of the owls by collecting pellets that the dogs detected. What I soon came to understand and what Max taught me from day one, is that a dog handler (or bounders as we call ourselves at Rogues), is much more than just a person in the field with the canine. Max understood in his very nature that for us to work together, we needed to have a bond. He was not a tool to be handled, and as such, I was not a dog handler. Rather, Max taught me what it was to become a bounder. For the methodology to work, we need to be bound to one another for a common purpose, to seek out data on cryptic species, yes, but also to be bound by love.

But I'm jumping ahead too soon. Let me back up. After accepting a job to be a detection dog handler, I packed up and left sunny New Mexico behind, arriving in a grey and overcast Washington. Upon arrival, I was immediately introduced to the pack of conservation dogs. I was greeted with raucous loud barking. "*Who is this intruder?!*", the dogs seemed to be shouting at me. Or maybe they were shouting *"Ball! Ball! Ball!*" as I scooted into the break-yard and was soon covered in their muddy pawprints, my ears ringing with piercing barks.

My impression after my hectic first day was that the job was not the best fit for me. I like cats. I like quiet. I like sitting alone and reading books in some tucked-away corner. These energetic dogs threw my world into a chaotic, noisy place and seemed more than I could handle. What had I gotten myself into?

I kept at it though. I had just moved across several states and I needed this job. Everyone told me that I was lucky to find work in this field.

I met Max a few days later – or rather, Max found me. I was working outside, helping to construct a fence, when I noticed a shadow. As I moved around the yard getting tools, a pointy-eared blue heeler who seemed more coyote or wolf than a dog was following my every move and stayed close to my side. I wondered if I smelled funny or if he wanted me to play ball with him. Being new to this work, I was definitely exuding more nervous energy rather than any, *"I'm fun, come hang out with me"* signals. Why was this dog attracted to me?

Max had been with the program for about a year before I arrived and was much loved by the lead instructor. The instructor doled out dog attention equally, though, and couldn't always be focused on Max. For reasons unbeknownst to me, Max chose me. Out of all the handlers in the program who had come and gone or were there currently, Max had attached himself to me. From that day forward, something shifted for me and I would do anything to be with him.

This meant that even if I was working with another dog, on another project, I still thought of Max. I stayed with the program long after my temporary four-month owl gig because I wanted to connect with Max, even if that meant volunteering. I guess it was sheer persistence but I finally got to work with Max. Our first project was searching for grizzly bear scat in Montana. It was a glorious assignment, despite the 3 am wake-up calls, trekking through dense devil's club (or Boudner's Bane, as we refer to it now), and avoiding any moose we came across, mainly because Max and I spent all of our time together. I admit that I cried when I had to say goodbye to him to go to another project without him. Then, when I was being furloughed during a slow spell, I begged the instructor to let me take Max with me for the winter. For months, we were vagabonds together, living out of my car and travelling the desert southwest until we could return in the spring for projects.

Max opened up something in me that I didn't know I was missing. He was my second half, my better half. He was all the things I wanted to be – a courageous, spontaneous, and playful counterbalance to my shy, serious, introversion. He was silly, patient, accepting, and understanding. And he was smart! He could find any scent in any terrain in any country – grizzly bear, spotted owl, marten, fisher, mink, lynx, bobcat, cougar, wolf, moose, tiger, leopard, and even orca whale scat throughout his career. His contributions to conservation science are immeasurable, but an untold, unsung song because the dogs who conduct this work are often not mentioned by name, just as 'scat dog' or 'detection dog' in research papers.

Max was something else, something beyond being my co-worker and my best friend, something so dear and special that I still haven't found the words to describe.

Max retired from fieldwork in 2019 after 10 years in the field. He was 14 years old. Soon thereafter, I learned that he had a rare cancer. What should have been his golden years, getting to chase balls whenever he wanted or sleeping by a cozy fire, we spent precious days travelling back and forth to the vet for chemotherapy treatments and blood work check-ups. Max had always been eager for an adventure and jumped into my car ready to go explore new scents. But he came to avoid the car, drawing back, ears flat, tail tucked. I would sit on the floor with him as he received IV treatments of invisible drugs that were supposed to help save him. Losing Max would be devastating. I had never felt alone until I realized I was losing Max.

It's autumn now and several months have passed since I said goodbye to Max. It happened on a cold spring morning, about the time COVID started to kick in, which meant, cruelly, that I was not allowed into the hospital with him. The vet, though, was so kind. He carried Max outside so we could be with him in his final moments.

I think I am still in denial that Max is really gone. Max was my world, my reason to be. He was my teacher, my friend, my mischief-maker, and my endangered species data finder. He taught me how to be a better human, how to listen, be patient, and laugh more, not take life so seriously. With him, I learned to appreciate all the tiny natural wonders the world has to offer: Look, a bug! What's that flower? Whose scat is this? What a big, beautiful tree over there! His absence leaves a hole in my heart and makes me realize now just how incomplete I was before he entered my life.

I wish I had had more time with Max. If I could go back in time and relive even the most challenging times – like how I initially struggled to read his subtle alerts, or how I felt like I was failing at being a good handler if we did not locate data – if it meant I got another moment with him, I would go back in a heartbeat.

I should not say that I am alone. I have worked with and fallen in love with many other detection dogs in my career. Scooby, a black lab bundle of love and cuddles, was Max's cohort and the three of us often went on projects together. He also recently retired, and I just learned that he has cancer too. Even though they came from different shelters and we all met late in life, I am always in awe about how dogs are so willing to accept newcomers into their pack. These days Scooby has helped me welcome in our newest pack member, Filson, another blue heeler mix. I am grateful to have them in my life and I know Max would say:

*"Get your head out of the sand and go play!"*

To all the *Lonely Conservationists* out there, a message from Max: You are not alone! Happy trails and happy tails to you from "rogue dogs," Max, Scooby, and Filson. Thank you for reading about Max.

## Update, 2021

In spring of 2020, not a year after saying goodbye to Max, I said goodbye to Scooby. His cancer caught up to him finally. He is buried next to Max in our home in Rice, Washington and I am glad they are near to one another. Whenever Filson and I are back from a field project, I visit with them and hope that they are having wonderful adventures together. To Maxi and Scooby, you are my heart.

# | 17 |

# Finding a purpose beneath the waves

## SARAH MUNRO-KENNEDY

I remember the day that I realised my purpose in life was to protect sea life. It was a few days after an event that occurred on December 26, 2004. I was 16 and on my first international scuba diving trip with my parents in Thailand. We had just gotten certified that summer and instantly fell in love, so we decided to take our Christmas vacation somewhere known for its diving. Little did we know what that vacation would hold for us.

We spent the first few days in Bangkok, visiting the most stunning temples covered in intricate tiling and gold, buying fresh fruits from the street vendors, and, of course, doing a little shopping. Then we made our way down to Patong in Phuket, where we spent several days at a hotel located a block from the ocean, immersing ourselves in the crystal-clear waters. The water was so warm that even when you were 20 meters down, you didn't need a wetsuit. *(Side note: we wore full wetsuits and I'm so grateful that we did. If you go to Ko Phi Phi, wear a full wetsuit! During our dives, I felt constant small stinging sensations on my face, hands, and feet. It turned out to be sea lice – not the kind that lives on fish and whales, but the kind that is actually jelly larvae. Thankfully, it wasn't very painful, but it definitely caught my attention).*

The sea life was amazing. There were so many fish darting around the reefs. It was like a rainbow swirl with large schools of brightly coloured fish dancing in the currents. We did a wreck dive and I saw my first sea turtle, and then we did a shallow dive and I found my first shark! I was looking for an octopus when I ducked beneath a coral ledge and saw her lying there. She was a small nurse shark, maybe three feet long, and I was obsessed. I hovered there for what felt like forever but was probably only five minutes, watching this shark rest. It was a moment that resonated with me and that I'll never forget.

After six days of diving, snorkelling, and relaxing on the beach, we were due to fly back home. Our flight was in the afternoon, so my mom and I had planned to go down to the beach and do some snorkelling before we left. Unfortunately *(yet, fortunately)*, I slept in. My mom came to my room at around 8 am to tell me that she and my dad were going to get breakfast. I decided to skip it and hop in the shower.

After my shower, I was busy blow-drying my hair when suddenly the power went out. I thought it was a bit strange and began to look around the room to figure out if it was just the blow dryer or the whole room that had blown. That's when I heard it. If you've ever been to a large waterfall, you'll know the sound I'm talking about. I walked out of my room and saw thousands of gallons of water rushing down the narrow street. I heard the cracking of tree branches as the water pushed past. And yet somehow, the island was completely silent.

A woman ran past me and I snapped out of my daze. I initially thought that a pipe had burst, but when I saw a truck being pushed back down the street, I knew that this was more than just a pipe. I realised that my parent's room was beneath sea level and their passports were in there, so I ran downstairs to gather their stuff. When I got halfway down the stairs, I stopped. The passageway was flooded almost halfway up the door. I stepped down into the water and opened the door. The water rushed in and soon the room was flooded up to my knees. I grabbed everything important that I could manage, ran it up to my room, and came back down to get their soaked suitcases. When I walked back into the room, I looked down. The colourful fish that I had just spent the previous day swimming with were now lying dead on the floor. I looked out my parent's window to the pool, saw destruction, and that's when it hit me; this was a wave – a, *really, really* big wave. I had heard of tsunamis, but like most people, I never thought that I would be involved in one.

My parents managed to find a way back to our rooms and found me, still trying to grasp the reality of what just happened. My mom is a registered nurse and she's done work in rural areas that have had natural disasters. She knew that if we didn't get to the airport immediately, we would be stuck in Thailand with limited resources, possibly for weeks. She told us to grab our valuables and we were going to find a way out. At this point, the water had receded to just around our ankles and it was much easier to move around. We began to make our way to the main road, walking down the narrow street that 20 minutes earlier had been easily four feet deep in water. As we were walking, we began to hear a rumbling sound growing louder and louder. We ran up a set of stairs to the roof of our hotel. When we looked down, the devastation and destruction were immense. There was debris everywhere, cars overturned and smashed, buildings destroyed, trees cracked in half, sand covering everything. There were several other people on the roof, all in just as much shock as we were. We all stood in silence as we watched the second wall of water forcing its way through the town, bringing down everything in its path.

The next half hour was a bit of a blur for me, but I remember helping my mom give first aid to those who needed it, then she told me, *"Let's go."* I didn't know where we were going, but I knew that she would keep me safe. As my parents and I made our way down the stairs, some of the people on the roof told us not to go. We took our chances and quickly ran down the stairs and towards the main road. We managed to find a taxi who had been up in the forest at the time of the waves and was now driving people up and down the hill. The driver wanted 100 USD to take us up the hill and my dad, who loves to haggle, stood there trying to negotiate with him. My mom and I climbed into the back seat and yelled at my dad to give the man the money and get into the car. He didn't see what we saw – a third wave coming. He jumped into the car and the driver sped off. The scariest part was that we had to drive towards the wave to get to the street that goes up the mountain. We skidded around the corner and began the climb up the mountainside. We were about 20 feet up the street when I looked back and watched as the water completely covered the main street that we had just been on.

We made it out safely with just a few cuts and bruises and considered ourselves extremely lucky. I struggled with sleeping for the next few days so I went to talk to a therapist about it. I was having what he called survivor's guilt. He told me that one way to overcome this was to find my purpose. And that's when I remembered all of the stunning fish I had swum with laying around my feet.

I began to think of all the corals, the turtles, the sharks, and the eels. I even thought about the tiny sea lice that stung my lips and cheeks. I wondered if any of these animals had survived this catastrophe. I couldn't find any information about it, and I was too young to travel back there myself to find out, and I felt somewhat defeated. Then I remembered the butterfly effect. The smallest flap of a butterfly's wing can cause a ripple effect that leads to huge changes. I decided that if I couldn't help the sea life in Thailand, I would help the sea life in my local waters in the hopes that a ripple will occur to help the rest of the sea life. And that was the day I found my purpose and became a sea life conservationist.

# | 18 |

# The empowering and saddening experience of being a conservationist

## SOPHIA NEIBLUM

To be honest, I wasn't sure whether I should write about my story. I'm in high school and haven't had nearly as much experience in conservation as many others. But I ultimately decided that my experience, although limited, is unique in some way.

My name is Sophia Neiblum, and I'm a 17-year-old living in Pennsylvania, USA. A few years ago, I began to read books about marine life, environmental science, and the climate crisis. Since then, I've read many books on these subjects: *Beyond Words* by Carl Safina, *The Soul of an Octopus* by Sy Montgomery, and *The Sixth Extinction* by Elizabeth Kolbert. These books, more than anything else, helped to spark my interest in the ocean. Throughout the years, reading has been a consistent force in my passion for nature and the environment.

When I was a freshman, I heard of a few people who had graduated a year early from high school, and the idea immediately captured my interest. Doing something exciting and meaningful in conservation, rather than spending my senior year in high school, sounded like the perfect idea for me. I'm not very extroverted, so I don't have much of a problem missing prom and other senior year events. Graduating early is uncommon in my high school, so it took quite a bit of negotiating with administrators and guidance counsellors. In the end, though, they were very accommodating. I'm grateful for their help and for the support and encouragement I've received from friends and family.

I graduated from high school in June 2021 and will be entering my gap year this fall. I'm participating in a Sea|mester study abroad program, where students live aboard a schooner for three months while learning how to sail and dive. I'll also be taking college classes in the marine sciences. I don't have my plans for the spring of 2022 figured out yet, but I'm currently looking for internships at local aquariums and nature centres, and I plan to work on the side to save for the future. I'm quite excited for the year ahead, and I'm very grateful for the ability to take a gap year.

I'm also well aware of the social aspects of graduating early and taking a gap year. I realize there will be days when I'll be doing difficult work while my peers go to parties and school dances. That I will most likely lose touch with people other than my close friends and family. That social media will become difficult for me to use, seeing how removed I am from my friends and my normal life. In that way, among others, I feel like a lonely conservationist. But these social aspects aren't deterring me—I'm confident that the experience will be worth it.

Being a conservationist is both empowering and saddening at my age. Participating in cleanups and climate action marches gives me hope for the future, but reading about the possibilities of climate breakdown brings me back to feeling somewhat hopeless. I led the Environmental Club at my school last year, and sometimes I couldn't help but feel as though we were never doing enough. I assume this relates to all conservationists. Our club did highway cleanups, planted trees, and tried to convince more people to recycle. I just think of the futility of those actions in the grand scheme of things, and it leaves me feeling quite sad. I need to remind myself during these moments that every small action does make a difference.

One way that I deal with climate anxiety and other worries is through writing. I took a creative writing class a few years ago, and it sparked my interest in writing poetry especially. Creating something from nothing, whether it's a poem or any other creative endeavour, is so therapeutic. When I feel overwhelmed about the state of the world, climate-wise or otherwise, I find that writing keeps me sane and helps me process the anxiety I feel.

Here's a poem I wrote in May of 2020, called 'Fuel' which was inspired by a quote from the 2014 book *All the Light We Cannot See,* by Anthony Doerr:

"*All of it is burning. Every memory he ever made. He thinks: the universe is full of fuel.*"

*Oil, coal, natural gas.*
*$CO_2$, methane, plastic.*
*The universe is full of fuel.*
*Solar, geothermal, wind.*
*Offsets, alternatives, progress.*
*The universe is full of fuel.*

*Greed, apathy, selfishness.*
*Hatred, ignorance, fear.*
*The universe is full of fuel.*

*Growth, learning, change.*
*Empathy, peace, acceptance.*
*The universe is full of fuel.*

Writing my story has allowed me to take a real look at my relationship with conservation and climate anxiety. I'm so glad I found the *Lonely Conservationists* community. Thank you so much for reading.

# | 19 |

# My mental health cost me my conservation career but helped me find myself

RHIAN

My mental health has defined my life. It lost me the conservation career I had worked so hard for, and yet, it has helped me grow as a person. I only ever tell people parts of my story when I feel overwhelmed, but never the whole thing.

So, here goes.

As a child, I didn't have a lot of stability. My dad was in the military and so, we moved every few years until I was 11. For every place we lived, I always went to a couple of schools as I could never fit in. The bullying would get so bad for me that I had preferred to move schools than stay where I was until we moved on again. It meant that my education was a bit patchy, to begin with, and my dad had to teach me to read and basic maths.

But I loved animals. Snakes and dolphins were the focus of my attention for several years, and I would read everything I could about them. And Africa. I was fascinated by the wildlife and culture of Africa and made it my mission to get there one day. My dream of working with wildlife was my driving focus to keep going and put up with the bullying until such time that I could escape, seeing as my parents had refused my repeated begging to be homeschooled. Even after settling down at one school once my dad left the military, the bullying didn't stop. I had to work very hard to get good grades and try and hide my knowledge gaps (when I asked in religious studies who Satan was at age 12, I didn't hear the end of it for a long time). I hated life.

End of 6th Form. I'd endured school and had got into my first-choice university course in zoology. I was excited to go, sure that I would meet like-minded people. But first, after leaving 6th Form, I got the chance to go on a biodiversity research expedition to South Africa funded by a youth development grant. This was my dream come true, I was so excited, and I wasn't disappointed. I fell in love with South Africa and the bush. The humbling experience of being surrounded by the megafauna of Africa made it feel like everything I'd endured to get to that point had been worth it. I'd found my calling. I was going to work in African ecology and live in the bush, I naively pronounced to myself. And I was about to leave home and the hell of school life.

Excitedly, I went to university with my life planned out. I met another student who'd just finished his field guiding qualification in South Africa – I was there! I ended up working two jobs non-stop over my holidays to save up the money, and eventually, I saved up £3,000 to go back to South Africa in two years. Yet, my mental health was taking another turn for the worse. I'd fallen into a group of friends, and one of them was a sociopath. This person ended up breaking every one of us over the course of our first year and by the end of that year, two of my friends tried to take their own lives.

I was a wreck. Panic attacks, depression, anxiety through the roof. And guilt. Guilt that I didn't try and save my friends. We eventually broke free of this person, but the damage stayed with me for years.

Finally, I got to go back to South Africa to study for my field guiding qualification. This became a healing process for me. I was back in the environment I loved, being treated with respect and feeling freed while living in the bush for a couple of months. This is my place, I thought. I was now qualified to work in the bush, and everything was worth enduring to be here and get my career in African conservation.

Despite my mental health still being very fragile and a lot of challenges with everything that had gone on, I went back and finished my final year of undergrad. I managed to get a free biodiversity research expedition to Indonesia and Malaysia straight out from university after working throughout my degree, and while I tell myself it was a great experience, I didn't love it as I'd loved South Africa.

After three months in Asia, I returned to my parents' house. I spent a year and a half trying to get a job in conservation and didn't get a single interview. The pressure I was putting on myself and the expectations of what I should have achieved, while others I knew were working, really got to me. As much as I'd hated school and struggled through my undergrad, I wanted my dream so much that I signed up to do a Masters in Conservation Biology in Cape Town. I had to wait a year to start, but in the meantime, I worked to save money for the fees and managed to get some volunteering experience back in South Africa to oversee the data collection for biodiversity surveys in the bush. It felt like I was getting back on track. But while I'd loved my position, it was marred with slight bitterness when I found out that I was the only member of staff not being paid, despite needing to have had my field guiding qualification to do my role, and I was working almost every job that everyone else was doing who was getting paid: guiding, lecturing, supervising. But I was getting experience, which was all that mattered.

Cape Town 2014: the start of my Masters. I was living in South Africa, studying what I loved, and I'd even managed to impress the course director with my African bird knowledge in week one – I was living the dream! But slowly, my mental health began deteriorating again.

The thing is, everything in South Africa is about race and skin colour. I started to feel so wrong being a white Westerner trying to help Africa save its wildlife. And the Masters' director loved pointing out when the African students did better than us Westerners. I was also spending all my time studying about how everything was dying, climate change, the COP conferences that weren't getting anywhere, and these were definitely not issues on the political agenda that meant anything would be solved anytime soon.

I felt overwhelming despair, not believing little me could do anything to save or change the world. This was long before the term 'eco-anxiety' came into existence and was openly talked about, but this is what I was experiencing acutely, and I felt entirely alone with it.

Then one night, a gun shoot-out happened outside my flat two minutes after I got home. If I'd been delayed by only two minutes, I could have been caught in it. The violent crime rate in South Africa is off the scale and, being a young woman, I was constantly on edge if I was on my own or in the dark. I felt guilty for feeling this way because this is a reality for people in South Africa. But after six months of slow mental breakdown, it was too much for me. I quit my dream, left my Masters and left South Africa.

Having walked away from everything I'd been working towards really broke me. The guilt for giving in, grieving the loss of my dreams and future. But I moved to a new city, got into a relationship (that would turn out to be over three years of psychological abuse before I could escape), and eventually decided that if I couldn't save the planet, I'll try and help people. I got a job in a mental health support service, and quickly rose from being the new person to running the service in about seven months! I had rock-bottom self-esteem but slowly, after some great mentoring, I was gaining confidence.

I was thrown into having to deal with behaviourally challenging people and assault, and this, sadly, was as much from the staff as the patients. But I found a new level of confidence, even while I was still suffering from my past and with an abusive partner in the background. I eventually quit after making a big mistake at work and, feeling guilty still for having left Cape Town, I got some freelance work as an ecologist.

The problem was I still hadn't dealt with my mental health. Spending hours alone with my thoughts in the dark on bat surveys, I mentally couldn't take having that much thinking time, and I left after a month. I carried on with social support work, helping vulnerable people deal with everything from mental health crises to debt and benefits. I was gaining so much confidence in myself and my ability to build relationships with people again. I realised that if I'd stayed in Cape Town for another six months to complete my Masters, I probably wouldn't have ever found this new me.

Yet, the guilt stayed, of having given up on my dreams. I wanted to return to my love for wildlife, but I was too scared to go for it in case I failed again. I eventually decided to go for another Masters, but this time, to take a more positive approach through sustainability to solve problems, rather than just learn about them. My new confidence made studying so easy. I had a new perspective on life, was loving what I was doing, and eventually turned my thesis into a published paper. I thought, "*OK, I can do academia again, it didn't break me.*"

I was very lucky and straight out from my Masters, I got a contract at a university to work on a project for the European Commission on climate change. I realised that I now had the mental resilience to deal with other peoples' petty temper tantrums (I'd dealt with physical assault and verbal abuse: an academic meltdown was nothing!) and that this was something people older and more experienced in life were struggling to deal with. I still have periods of depression and the constant imposter syndrome, but I managed to not let it overwhelm me as it once might have.

I've now decided that I want to have another go at trying to break into the conservation field again. But now, the problem is that my CV looks all over the place to an outsider who hasn't been on my mental health journey. I've had my motivation and integrity questioned so ruthlessly during one interview that, despite my initial reasonable response that I'd left my Masters in Cape Town due to health reasons, I got to the point where I broke down in tears and had to explain that I'd had a breakdown to make them back off and understand why my CV is what it is.

Despite that, I do overall feel more mentally resilient, mature and capable of working in wildlife conservation than if I'd stayed that extra six months in Cape Town and have a piece of paper to say that I have *the right* Masters in Conservation Biology. I fear that my history is still holding me back though. But right now, I'm carrying on with my applications hoping that I might one day return to the conservation career I always wanted.

# | 20 |

# Getting away to feel at home

## ANGUS HAMILTON

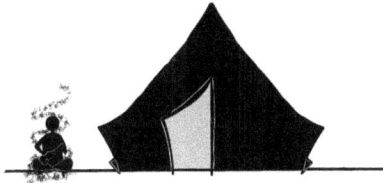

My conservation journey is a tale of two parts. It started when I was very young, as most of us have done, watching *The Crocodile Hunter* and wanting to be everything that Steve Irwin was, and taking in everything that he stood for. I spent hours and hours playing in the back garden pretending to be him. I'd gently hold plastic lizards, wrangle plastic snakes, and wrestle plastic crocodiles (all while family friends, or 'zoo guests' watched on). I also spent almost every weekend at the zoo, watching and learning about all of the animals I could.

Over the years, I began to see other interests and focuses start to take priority as I discovered that learning about wildlife was almost impossible within the school system. I discovered that the traditional sciences (chemistry, physics, and even biology the way that it was taught) did not completely agree with me. History, politics, and English became my main interests. My passion for animals was still there, but I had no avenue to pursue it at the time. I was too busy playing sports and trying to write essays.

I was able to convince myself that this was all OK and what I wanted. That is until I started getting towards the end of my university degree. My group of friends was full of people studying law, business, and politics. All of these people seemed to know exactly what they wanted to do and had a path towards it. I, on the other hand, felt the complete opposite. I was studying international relations, but I could feel that this was not truly what I wanted (or perhaps even suited). The more I came to understand that arena, the more of a misfit I felt. I felt lost, and like I'd lost myself as well.

When I graduated in 2016, I still was totally in limbo. I was working full time in retail, which was unsurprisingly an unfulfilling job. However, it wasn't long until everything changed.

A friend messaged, asking if I'd be interested in backpacking around South America. I don't know if I could have said 'yes' any quicker. But I also saw it as an opportunity to try out something that might provide an insight into something I'd like to do in the long term. I recalled a volunteer program I'd seen once, in Madagascar. Behavioural studies of lemurs, trekking through the forest looking for reptiles and amphibians, and point count surveys for birds. What did I have to lose?

Landing in Madagascar, with a plan of five weeks there, before six months in South America (it wasn't the most economical plan at the time, I have to admit), I had no idea what was in store for me. Driving from the airport to get on a boat to the small island I'd be working on, a flash of bright green with a splash of red stood out. My very first panther chameleon *(Furcifer pardalis)*. I was well and truly hooked.

It was several days later though, that I knew I was truly onto something. Trekking up the longest and most difficult trek to the top of the island, to conduct plot surveys turned out to have been everything I could have hoped for. Snakes, geckos, chameleons: I was in heaven, all of that passion for wildlife and nature that had been hidden for so long was starting to re-emerge.

We hiked up (and up and up) and got to our first plot survey site. I'd been in Madagascar for three days and had been trying to learn all of the species. It wasn't easy, but I was also enjoying discovering how to ID the different wildlife found here. We set up the plot, and began the survey, actively searching through the leaf litter to see what we could find. After about two minutes, as my fingers moved a piece of the leaf litter, a tiny little body fell into the hole that I had left. I looked down in shock at this little grey body as it wriggled and squirmed, almost worm-like. But it was no worm. Instead, it was one of the leaf chameleons endemic to Madagascar, and the second smallest lizard species in the world, *Brookesia minima*.

That first month in Madagascar changed everything. My friend had to cancel their South American plans, and I was offered a job as I apparently had a bit of a knack for this whole wildlife and working with volunteers thing. Five months later, I left Madagascar with the most important thing I could have had: my passion had flooded back into me, stronger than anything I had felt before. But I also realised that I loved to share my passion with others. One of the best parts of those months in Madagascar was teaching volunteers that arrived about all of the incredible animals found on the island, and what made them so special. It was incredibly rewarding seeing the moment that they went from not caring about that bright green gecko that was just lying around, to seeing the realisation in their eyes as they noticed that actually, these were pretty cool animals too.

There's a quote from Steve Irwin that has stuck with me since I was quite young, but has become increasingly relevant since I first arrived in Madagascar:

*"If we can teach people about wildlife, they will be touched. Share my wildlife with me. Because humans want to save things that they love."*

This idea that people want to save things that they love, but they can't love what they don't know really inspired me to try to take my passion further. So I started Life Gone Wild, a website to share articles that I wrote about some of the weird and wonderful wildlife that we share the planet with, trying to ensure that more positive stories were told about these species, but with a focus on the efforts that people from around the world were trying to implement to save them from extinction. At the back of my mind, the hope was to one day be able to produce videos to add to this platform, to engage people even further, but there was no way I could do videos. I didn't know enough, and who would want to watch them anyway?

A year later, a competition asked for a video entry, a minute long, and my favourite science fact. While I didn't win, just entering was enough to start me down the path to video. Forcing myself to just give it a go and see what happens meant that I had to put aside my insecurities and uncertainty to just do it! And I couldn't be happier that I forced myself to enter.

Somehow, I've been creating videos ever since! Most recently, I've been producing those while conducting herpetofauna research in the rainforests of Peninsular Malaysia. Sometimes the videos are intermittent, given life in the forest and limited Wi-Fi capabilities, but they go up eventually.

Who knows what the future holds though? I barely know where I'll be in six months, considering so few of my plans end up working out how they were meant to.

I'm still a little petrified of putting myself out there, standing in front of the camera, and then posting that online for the world to see. But at the end of the day, it's worth it. It has to be. The wildlife that we are all working to save doesn't have time for my insecurities to hold me back any longer.

# | 21 |

# For the elephants

## RACHAEL GROSS

You know when you're a kid and you get asked, *"What do you want to be when you grow up?"* The answer is usually something you've seen in books or on TV – a doctor, an astronaut, a lawyer. When you *grow up*, all that changes. Life happens, it becomes more real, and you often end up in something more specialised but probably not where you thought you'd be as a kid. I was different and really lucky.

I've known since the beginning that I wanted to save the elephants and that I wanted to live and breathe Africa. I couldn't tell you why, but while most kids were reading fairy tales and watching cartoons, I refused to read anything except animal reference books or watch anything except documentaries. Though, on reflection, I feel like most conservationists felt this way. That passion, from the beginning, is what sets us apart. My idol, Jane Goodall, summarises it well in her documentary *Jane*:

*"I was typically a man; I went on adventures. Probably because at the time I wanted to do things which men did, and women didn't."*

Going to Africa, living with animals - that's all I ever thought about. I wanted to come as close to talking to animals as I could, to be like Doctor Doolittle.

I wanted to move among them without fear, like Tarzan. Among the huge, gnarled, and ancient trees, and the little streams chuckling their way through rocky pathways to the lake. The birds. The insects. Since I was eight or nine years old, I had dreamed of being in Africa, of living in the bush among wild animals."

My goal was clear but my journey was not smooth.

I came from a poor, dysfunctional family. My rural Australian high school experience was gruelling – I hated it. My dad was an alcoholic who drank himself into disability. For most of my life, he lived in a nursing home. He lost his short term memory, so after a couple of years as I grew up, he stopped being able to recognise me. My sister attempted suicide for the first time when I was 11 years old. My brother was both physically and verbally abusive, which has left me with plenty of trauma. They are both a lot older too, so our relationships were a rollercoaster. My mum was and has always been a superwoman; she did the best she could with nothing. On top of the trademark insecurities of high school – being weird, chubby, and having both glasses and braces - I was bullied mercilessly and it wasn't a good time. Despite this, I got into the best university in Australia, and I followed (forged?) my path.

The only part of me that didn't change, that no one could bully or take from me, was how much I loved wildlife. My passion for a career in it never faltered once, even when everything around me was unsteady.

Moving away was an opportunity to start a new life, and I took it. I didn't really fully come into my own for a few years; it takes a while to shed insecurities and grow. I flicked my way through biology, earth science, environmental science, and landed on biodiversity conservation. Inevitably, the tests continued. My sister attempted suicide again in 2014, my dad passed away from lung cancer after a one-year battle in 2016, and I was prompted into realising in 2017 that I had been drugged and raped at 17 years old. The silver lining was that I came across an excellent counsellor in 2015, who I've seen ever since.

I ended up doing an honours project about two of my most impassioned topics: elephants and climate change. I couldn't believe my luck! I lived in South Africa for three months, which is still the highlight of my life so far. Unfortunately, I had a truly awful supervisor, and she cost me a lot. When I got back from South Africa, I was diagnosed with High Functioning Anxiety and depression. But the person I was in Africa was the best version of myself that I've ever been. I was a strong, bold and respected woman who spent every day out among the elephants – was that not the dream?! Even though I've looked back on this time to reflect on mistakes I made, mostly my involvement in the perpetuation of neo-colonial oppression of local people, I still miss it. Just with the knowledge that I'll do it differently the next time.

When I somehow finished honours, I was done with academia. My secondary supervisor made it clear to me that he wanted me to do a PhD, but I couldn't. I'd been at university for five years; honours had nearly killed me. So I took a year off to work in a stable job that I was offered in science recruitment and outreach for the university. I thought I was safe, but unexpectedly, I was thrown into a deep depressive episode.

Suicidal thoughts were creeping back; I was sleeping for 10-11 hours a day. I didn't see my friends or family. I barely left the house and I was self-harming again. I started seeing a psychologist and while she didn't work particularly well for me, I was diagnosed with depression and put on medication. It took a few months to stabilise, and I've put on weight, but I finally feel like I'm coming back to normal.

In the words of Elizabeth Warren – *Nevertheless, she persisted.*

Now I feel like there's so much more to me that I'm not afraid to back down from any more; feminism, climate change, intersectionality, Indigenous issues, mental health – you name it and I'm probably angry and outspoken about it. I've learnt to never apologise for being a powerful fucking woman.

I'm now doing a PhD. After honours, my second supervisor asked me, *"Do you still love elephants?"* and I laughed, *"Of course, they're the only thing that got me through." "Then you need to do a PhD. You've got what it takes and more."* Now I study a beautiful intersection between elephants, climate change, and community-based management, and I couldn't be happier (except for that millennial burnout about the world being on fire).

I know living in Africa is maybe a year or two down the track, but now I get paid to think about elephants every single day. How unbelievably lucky am I to be able to say that? I'm doing better than I ever thought I could be, and one day I'll be a 'Doctor of Elephants'.

To be honest, I adore all animals, but it has always been elephants. Spending time with them changed my life for the better; being in the bush and sharing their space changed me for the better, and being in Africa changed everything for the better. It's my safe and happy place. I'm lucky to have that on a pedestal to strive and work towards always. I'm lucky to have been diagnosed early, and I'm lucky to have accepted it. I know PhDs are riddled with tales of stress and isolation, but I have never once felt isolated. I'm surrounded by the most incredible people and at the end of the day, if I'm stressed – I'm probably stressed about elephants. What a privilege!

Anxiety still creeps into my life every day. Little things are difficult. I still need a lot of sleep and have become flakier. I even get anxious about the idea of going back to Africa, even though it's what I want more than anything. But now I know when those thoughts are anxiety and when the feelings of giving up are depression (or that I need some sleep), and I can usually wait until it passes or step over them towards everything that I've earned – a happy life with beautiful people and full to the brim with my elephants. I work in science communication; I inspire people about the world of conservation; I contribute to the conservation of elephants – I don't know how I got here, but I'm mighty glad I made it.

In summary, a poem by Mark Anthony:

> *And one day*
> *she discovered*
> *that she was fierce,*
> *and strong,*
> *and full of fire,*
> *and that not even*
> *she could*
> *hold herself back*
> *because her passion*
> *burned brighter*
> *than her fears.*

While I can't pinpoint the exact day that I discovered my fire, I know it'll always be there burning brightly. While my journey to being a conservation biologist has been far from smooth, I believe that what doesn't kill you makes you stronger. I've accepted that it's OK not to be OK. I'm now proud, instead of scared, to tell my story. And at the end of the day, every day, it's all for the elephants.

## Update: 2021

While every day is still all for the elephants, I guess it's appropriate to update my story to fit in the world of COVID-19. It's interesting reading back on this, and reflecting on the world I was living in. I was just six months into my PhD, and I can see the genuine joy and feelings of security that I had.

I am still doing a PhD: in fact, I'm halfway through. I have one paper under re-review and one ready to submit, but everything has changed. The cancellation of my fieldwork was, despite everything, one of the hardest things I've ever dealt with in my life. Mostly because my life was entirely dedicated to it: all my eggs were in one basket and I made a lot of massive sacrifices for it to happen. I did a PhD for that fieldwork, that's huge. And it's gone.

Soon after, I lost all my jobs because of the COVID-19 recession and I was living off my stipend which is half the minimum wage. I went from feeling like I was living my best life, and that things were finally coming together, to where I am now – uncertainty but sureness in knowing that this isn't what I ever wanted or planned. I still work on elephants, but with distance comes a drain on motivation. I won't get the skills I thought I would get, meet people I thought I would meet or see my elephants. I won't get to be the person I was when I was doing fieldwork in South Africa.

It's all gone.

And I feel bad for feeling this way when people have lost so much, including their lives. But the last thing we need to do is gatekeep our own feelings during such a traumatic and tumultuous time. I really have been struggling, sometimes with reprieve, but largely without.

One silver lining is that I have been able to throw myself into my work and passion for decolonisation, and the opportunity to reshape my project has been very conducive to this. My relationships have gotten better and I feel closer and more connected to many of my friends who are far away, but I still feel trapped in a life I never wanted without a foreseeable way out. One of the worst problems with anxiety is the inability to accept uncertainty, and uncertainty is all the world can offer at the moment. For all of us.

As the conservation industry keeps taking economic blow after blow, I need to reconsider my future. My future in the IUCN (my dream), my future living and working overseas and my future with elephants. When I think of the future, I struggle to think past the next few weeks at a time, but in 18 months, my PhD will be due and I will have to decide where to go next.

My time and investment in decolonisation, and support from my research circle, has planted a seed for me. For the first time in my life, I'm considering staying in Australia and trying to work in that area to combine my interests in conservation and decolonisation. If nothing else, the pandemic has given us time to think (and stew and fester) on our future. I don't know anyone whose future hasn't been altered by world events in the last year, but for once, it feels like we truly are all suffering together.

As conservationists, we want to save the world. We want to fight for it, we want to change it and we want to improve it. The last 18 months have shown us that to do that, we need to work together, which means fighting for equality. I have a huge poster in my office that reads:

*"There is no environmental justice without social justice"*

It couldn't be more true. That's something we all need to internalise as we move forward, there's no future for conservation without fixing the systemic problems first.

But as my hope waxes and wanes, I still always look back to my elephants. My passion and love for them have gotten me through so much in my life, and I need to draw from that over and over again to handle the curveballs. We need to be kind to ourselves, and to those around us, and hope that it's enough to save ourselves so we can save the world.

# | 22 |

# (Who) to be and not to be

## PHALGUNI RANJAN

My journey to the field of conservation had a pretty clichéd begin-ning, now that I think about it. The classic marine biologist starter pack had some really simple components back then. Take an enthusiastic 11-year-old, throw in some Animal Planet, a devout love for water and animals, the desire to do something, and blend it all. For someone who had never seen the sea, dolphins or reef fish, the prospect of studying them was a huge deal.

Interestingly, as I grew older and with each step that I took towards my goal, I realised (almost always belatedly) that I'd thoroughly loved and enjoyed other aspects of the journey too. Aspects that weren't limited just to the marine realm or indeed, the concept of conservation as we understood it then. That neon sign was broken (*facepalm*).

I went on to complete my masters in marine biology, and landed an internship, and then a job. It was less than a couple of years after getting into research that I began to feel the disconnect. I was baffled – wasn't I doing what I wanted to do? Didn't I like my work? *"Yes, but....."* was the answer my mind threw back at me every single time. One year, a breakdown, and a short career break later, the sentence started completing itself.

*"Yes, but I feel I'll make more of a difference in a different capacity."*
*"Yes, but I feel I'm doing a half-assed job in research and I feel stagnant."*
*"Yes, but....no!"*

A year later, I had a new job handling bits of conservation communications, wildlife awareness and related content creation, and new insecurities, and I was quite adrift at sea. I asked myself the same questions again and got the same *"Yes, but...."*, but thankfully, it wasn't long before the rest came through.

*"Yes, but I feel like an imposter."*
*"Yes, but can I even call myself a marine biologist anymore?"*
*"Yes, but will anyone take me seriously anymore, and will I hit a ceiling too soon?"*

I was stuck between two warring identities- who I thought I should be and who I wanted to be. I had wanted to be a marine biologist since I was 11 years old, and I now wanted to do something more creative, and more people-connected. Something that I felt was more in line with my skills and aptitude.

Luckily, I had realised within just a couple of years of working that research wasn't what I wanted to do. Despite the early switch, the identity crisis and existential crisis hit home hard. Somehow, I felt extremely guilty and worse, that something was wrong with me. I'd worked hard to get there, studied the right courses, gotten the relevant degrees, and won a gold medal, only to want to change my line of work? Who even would I be anymore? Wasn't I letting myself down? Wait, I was letting others down – that was worse, wasn't it? I was convinced I was letting someone down.

It took me months (and talks with many kind folks) to realise I could wear multiple hats in terms of identities, and that I would be a marine biologist irrespective of whether I was active in research or on the creative conservation or communications side of things. I do still rely heavily on my knowledge of the systems, of all the concepts I've learnt and implemented, and my research experience, however short it may have been. My art is still inspired by marine life to a large extent, and someday, I want to carve out a niche for myself in a capacity closely connected with that world.

The portfolio I now handle is diverse and challenging, and I'm surprised (and relieved) to find that I quite enjoy it and find it fulfilling. My acceptance of my change was as incomplete as saying I'm a jack of all trades, and construing it to be nothing really great. Thankfully, not anymore. Do you know the entire saying?

> *"A jack of all trades is a master of none, but oftentimes better than a master of one."*

Cue, self-image boost. A Jack (or Jill, if you will) of all trades is definitely what I am. And I'm learning to own it. I am at my creative best at work and outside, now. As a hobby, I create art that incorporates elements of different art forms and wildlife to create fusion pieces, many of which haven't yet seen the light of day but, oh well. At work, I've never been more motivated to creatively portray a conservation success story and illustrate some powerful figures that popped out at me from the data.

I now accept I'm a package deal and not just one label or title. That I wear many hats and change them as per the requirement or occasion. For the longest time, I felt like an imposter and sometimes on bad days, I still do. But then again, don't we all? Personally, I have yet to meet someone who doesn't have the same doubts or fears or insecurities, and I realised I wasn't alone only once I started talking to others. It has helped widen my horizon, and normalise the doubts and insecurities a whole lot of us deal with every day. What was once a huge, identity-crushing thing for me, became something not unique to me and in the process of normalising it, it lost its crippling hold over me.

It is true that we care far too much about what others think, but being part of a society and for better or for worse, sort of bound to the norms of the circles we run in, it takes a lot to fight it. We get caught up in titles and labels that we take to define us with such absoluteness that we don't leave room for ourselves to transcend boundaries and be something new.

I have come to realise that in this fast-paced, overly competitive world, sometimes we need to hit pause, assess, and share a little positivity, and that has been the motivation behind my art. To create something beautiful and positive from the negativity and turmoil within.

Conversations, discussions, and just saying it aloud helped me so much and I want to do that for others. Down days are normal. Doubting your work at times is normal. Feeling lost is normal. Insecurities are normal.

All of this is normal. But the key is to take time to acknowledge it, accept it, reach out, talk about it, and then bounce back to be the spectacular person you're meant to be!

# | 23 |

## Zero waste diaries

### TANYA JASWAL

As a child, I was always interested in wildlife. I loved the birds, animals and insects around me. I was very comfortable around them. To push my interest a little forward, I joined conservation organisations and gained knowledge about birds: how to identify birds by their call, by their size and shapes. Slowly, I became more sensitive towards animals and birds. Every weekend since, I visit nearby parks, sanctuaries and gardens to observe and document birds and insects, to learn more about them.

While nature was bright and beautiful, there was always a flip side there that disturbed me: waste. Whenever I used to go to my terrace, I would always wonder why kites were flying above an odd mountain in the distance. Someone told me that it is not a mountain, but a heap of garbage, clearly visible from my terrace. I was shocked but grew interested in that mountain so much that it made me take up the journey of urban waste management.

I visited Delhi's largest landfill (Ghazipur Landfill) for my research. I was on the heap of waste. Dogs were barking at me, black kites were flying so near that it looked as if they would scratch me. Big rats were running here and there in search of food. In the landfill, I saw kitchen waste, clothes and the enemy of the environment that is plastic. Everywhere, there was dust, a terrible stinking smell and the heap of waste was trembling below our feet. There were waste pickers who were working without any protective measures.

I felt disgusted and that was the turning point of my life. That moment I decided that whatever it took, I will devote my whole life to managing the problem of waste.

I started looking for internships and volunteering opportunities in different sectors around waste management. Beginning with solving issues of waste management in Himachal Pradesh (a state in India). Then working again in different parts of Delhi. Currently, I am working as a social designer and field coordinator at a company, on an ongoing project. Under this project, we have made two temples of East Delhi convert to zero waste by composting the flowers and organic waste, with the help of two drums installed by the organisation.

Earlier the flowers used for religious offerings would end up in the river and in the landfill but now, the compost is created from the flowers and is distributed free to the worshippers and the temple committee.

Recently in Holi (a festival of colours where people apply colours on each other and play with water) under the Su-dhara project, we trained some women in the community to make organic colours with flowers which were offered in the temple as organic *gulaal* (coloured powder). The ladies packaged the organic colours in small cloth bags (zero-waste packaging) made out of religious offerings of cloth to the temple. We also trained over 50 community members to compost their organic waste at home, a practice they have eagerly continued.

I also started composting my organic waste into wonderful compost and started living a zero-waste lifestyle. Living sustainably and sharing my knowledge, experience and passion with others has led me to create my zero-waste Instagram page, where I talk about composting and how to live a zero-waste lifestyle. I also portray the stories of waste pickers who face daily challenges working in a harsh environment.

During my research on waste management, I used to tell my friends and neighbours that I am working on waste management. People always told me to find some other job rather than work on waste, as it's the most disgusting job, according to them. Some of the comments were *"You will be surrounded with waste!"*, *"Why are you even concerned about waste? Let the government handle it"*. This wasn't enough. There was a time when I was conversing with our street's waste picker, and when I left, a policeman beat him up, telling him that he couldn't talk to a person from a "big house". After that, it took me around two to three months to make that waste picker comfortable again. Everyone demotivated me to work on waste. It's hard to change the behaviour of the people because people are resistant to any change.

There was another lady who was throwing her waste bag from the third floor of her building. I was filming her, and she called the police. Then I understood that waste management is something that people don't want to talk about or want to involve themselves in. I understood that waste is not just about waste; it is about identity, illegality, perceived illegality, and the insensitivity of the upper-class people who throw their waste from the top of their buildings. It is not just about segregation and composting. It is about human rights, human dignity and human safety.

When the whole world was battling with the pandemic and the whole country was under lockdown, the waste pickers were working. They were segregating the biomedical waste with their bare hands. The work that the waste pickers did was essential yet highly dangerous and thankless work.

*"We don't need a handful of people doing zero waste perfectly. We need millions of people doing it imperfectly"* - Anne-Marie Bonneau.

I would go one step further and say:

*"We don't need a handful of people composting their organic waste or living a zero-waste lifestyle perfectly, absolutely not, we need millions of people doing it imperfectly."*

We need more people to be aware of the waste management issue as it is something we generate and we should have a way to responsibly dispose of it. The issue of waste is not only a human concern but an ecological one as well. Some animals consume plastic unknowingly from the heaps of garbage. The bits of plastic consumed, build up in their internal organs and make it difficult for animals to eat. The same happens with the birds. So it is really important to spread awareness on waste management issues around the globe. Unsegregated waste from a landfill leads to toxins leaking into the groundwater too, eventually entering our food cycle and causing health issues in the long run.

Being more conscious about ethical waste disposal is vital to our survival as well as the health of our environment. We must educate ourselves and begin the small steps towards an eco-conscious future.

# | 24 |

# Born this way

## ROXANNE

### *The Early Years*

My passion for conservation, like many, has its roots in early child-hood. It was watching the well-known presenters like Steve Irwin and David Attenborough, but before Animal Planet was reality show-cen-tric, there were others I enjoyed and learned from. They were reptile enthusiast and photographer Austin Stevens; biologist Jeff Corwin; and zoologist Nigel Marven. I was also a frequent visitor to zoos, wildlife parks and museums.

I wasn't a child of the outdoors who camped and trekked the rainforest of far north Queensland where I grew up. In fact, I was seriously ill with a Chiari Malformation: a rare congenital defect of the skull compressing the cerebellum and brain stem into the spinal canal. The statistics are unclear, but it typically occurs in 1/1000 births.

The condition restricts the flow of spinal fluid and impacts the automatic functions for life, such as breathing, temperature regulation, motor control and swallowing. An excess of spinal fluid in the brain (Hydrocephalus), a tethered spinal cord and a spinal cord cyst also develop in patients. Brain stem decompression, spinal cord and shunt insertion surgeries until the age of five prevented significant brain damage, severe disability and other life-threatening conditions. After my last MRI at 11 years of age, I haven't received specialist treatment as I simply had enough of it all, was stable and if anything, more procedures would've been unhelpful.

What remains is the syrinx as it was too complex and risky to treat. Because of that and other residual symptoms, I have a low fitness level and problems with balance and coordination traversing over loose rocks and steep rises.

Engaging in environmental science and conservation through film, zoo visits and reading wildlife and palaeontology books helped me in these hard times and created a good knowledge base well before I was taught formally in senior high school and university.

## University and Professional Conservation

Studying environmental science at university and work experience in the sector brought the childhood dream to fruition. And to my surprise, my interests in conservation are more general. It started from a zoological leaning, but from studying extension subjects and working in the sector, I enjoy the social and humanistic aspects of conservation and working with people.

I did my undergraduate degree at a regional university in northeast New South Wales (NSW), studying mostly online with practical classes and field camps. Some of my favourite experiences were fauna surveys in Richmond Range National Park, an excursion through the national parks of south-east Queensland and northern NSW and engaging in discourses of natural resource issues and human impacts. Most of all, it was enriching to be taught by those I looked up to and socialise with other like-minded students.

While studying, I got straight into working in the conservation sector. My first job, at 19 years old, was as a Green Army Participant: one in a group of young adults working on conservation projects in a government youth employment program. We worked in local national parks and reserves controlling weeds and planting native seedlings with local Landcare groups. I enjoyed the work (including the physical), the people and the application of my first year of study.

After five months of working in the Green Army, I volunteered with the NSW National Parks and Wildlife Service for nearly four years. My main role was data entry for their pest management activities, but as the rangers knew about my career goals, they would let me go out in the field with them. I did flying fox, vegetation and shorebird surveys, as well as experience different aspects of protected area management. Like at university, the rangers were my core role models and friends.

I now work as a project officer with my local Landcare organisation. Despite COVID-19 impacting normal operations for a time, it's a great job after graduation. I've written for funding applications, attended committee and inter-agency meetings, and seen conservation dogs at work for the first time.

## The Challenges

Surgery attempted to reduce the impact of my Chiari but because of the physical limitations, it's hard to find the right career path in conservation. National park rangers and field officers require a high fitness level to work in very remote locations, fight bushfires and perform manual tasks. The same goes for field ecologists and ecological restoration practitioners. While I've managed myself as best as possible during university and working in the sector, it doesn't mean I can independently fulfil some of the conventional internships and entry-level jobs in conservation.

Unlike other disabilities and health conditions, specific supports for mine in the workplace are unclear. Field sites differ and it isn't always practical. Therefore, I may decline in participating in an activity. Then there's the situation where it doesn't impact my ability to work, like an office environment. While open to this, I'd rather see practical outcomes in my work and be outdoors on occasion. This fluidity of factors is frustrating seeking help from disability employment services and articulating my situation to an employer.

Other disadvantages I've experienced include living in a country town where the main industries are retail, hospitality, health and social services, corrective services, and construction; the limited scope for youth, unemployment and ineligibility for disability employment services and income support as a full-time student; and disconnection from organisations in regional and major cities.

There's also resistance and ambivalence on my part. I've lived a mostly normal life since my early teens, where my health wasn't an issue. I did very well at school and university, learned to drive and joined the workforce. Then there's consideration of the clinical aspects of my condition. For a while, I resisted engaging a specialist as I didn't feel unwell and wasn't comfortable with medical professionals.

## Commitment to Positive Actions
### Physical & Mental Health

I haven't seen a specialist in over a decade as it was thought my Chiari is unlikely to become worse and I'm mostly healthy. However, there are still symptoms to manage if wanting to be in the field occasionally. And something could happen later in life. I'm at an age and education level where I'm genuinely intrigued by neuroscience and see the practicality of an expert to connect me to services and resources. I'll soon work with a neurologist to monitor my condition and assist me in living and working more independently.

Long periods of reflection and conversations with loved ones have also encouraged me to consider the psychological impact of my life experiences. While treatment was so long ago, it's possible the trauma and my life since then likely contribute to how I think, feel and behave as an adult. I'm working with a psychologist to better understand this, manage mental health and be more resilient.

I'm also implementing things at a smaller scale, such as undergoing physiotherapy and working with an employment coach to access supports and adjustments for work, assist with further study, connect me to organisations and learn skills to be more independent in my field.

## Education & Experience

Because my work at Landcare mostly is casual, there's room in my life for exploring other things in the conservation field.

After graduating from university, I studied a project management qualification at TAFE NSW (a vocational training and education platform), where I got enthused by stakeholder engagement and human resources. There's also postgraduate study to consider, possibly specialising in wildlife management and zoology.

I also intend to volunteer wherever possible. Close to home in northern NSW, I'm preparing to join a large koala bushfire recovery project with an organisation lead by a university alumnus I admire. Elsewhere may be frustrated by COVID-19 travel restrictions but will happen eventually.

The aim is to try different things and meet different people, where valuable opportunities, learning and relationships will result.

## A Final Note

I understand chronic illness or disability is an intensely private issue for some as there is a fear of judgement or not being supported. It's something I'm still coming to grips with. That said, how can people in your field help when you don't say what's difficult? There are great, supportive people out there when given the chance.

Plus, fortune favours the bold. Continue to learn about your industry, try different interests and jobs, test your limits, and proactively approach organisations to connect with like-minded people not for just employment, to help develop you.

May this story provide consolation, optimism and options for those in conservation experiencing complex health conditions. More importantly, may it reveal an area of inequality not widely spoken and supported in the conservation industry.

Thank you to Ian and Warrick, my childhood neurologist and neurosurgeon who looked after me and my family all those years ago. I wouldn't have lived this amazing life without your help.

# | 25 |

# Ode to a curious life

## JACK O'CONNOR

Just north of one Port Phillip Bay,
water and eyes an ocean hew;
a newborn child took breath that day,
his life had begun new.

Jack was his name, of Irish line,
before a century's debut;
and so we start our steep incline,
into a world fresh and new.

In youth, obsessed with plants and birds,

endless novels to purview;
'Velociraptor' recorded in first words,
cataloguing species ancient and new.

Though social tact was... sadly missed,
into curiosity I grew;
mind set on palaeontologist,
parents set on pathways new.

See, I lived amongst a clan,
looked forward to fields new;
to them, fossils doom career plan,
for stability, choose something new.

Architecture sang of flair, of mind,
a job which could accrue;
so from there out the choice behind,
would nature stage a mental coup?

Walls of empty canvas boxed me in,
as a cage could surround you;
past building sketches came sound thin,
toward it, my attention drew.

It spoke of colour, of shades, of light,
a hope I could strive to;
pushing past the shades of white,
with renewed strength, I drew.

A canvas is a spotless board,

*creation limitations few;*
*and nature's inspiration soared,*
*open, the curtains drew.*

*Clouds danced from my frenzied keys,*
*whilst digital mountains grew;*
*graphic design I found with ease,*
*inward, the breath I drew.*

*I pioneered my PowerPoint art,*
*its boundaries I slew;*
*it drew me back to the start,*
*for nature, for life, I drew.*

*Now drawn back to life extinct,*
*I summoned all I knew;*
*to reach my course I hardly blinked,*
*and presto! Now uni to go through.*

*I fought hard to convince my kin,*
*pushed my scores to ensue;*
*to learn each phylum, limb, and fin,*
*and reach the point I wanted to.*

*Global Challenges was now my course,*
*a unique science avenue;*
*but the imposter wave's a mighty force,*
*a wall I'd run into.*

*I tutored classes, devised promotion,*

with talented peers to live up to;
I volunteered, scanned the ocean,
searching fields to break into.

SciComm brought me to the stage,
my creative side felt déjà vu;
I finally started to turn the page,
a stable road I could look to.

Study flew me 'across the ditch',
a Pūkaha internship held me true;
then a Malaysia exchange was my pitch,
avian linguistics I strove to.

Nine months were planned for over there,
I ended up with weeks two;
fate seemed to not be playing fair,
with a virus worse than any flu.

Trapped overseas as the world shut down,
bird research? Off it flew;
here settled a constant frown,
as I struggled with what's true.

My network helped us to escape,
I helped the students through;
I spoke our way through miles of tape,
then a flight home heading true.

That year would spiral, lost its shine,

trapped inside like fly to glue;
my mental health would soon decline,
couldn't tell the false from true.

Now as we open, I lose reserve,
growing to nature like bamboo;
deciding which I should conserve,
including me... had to be true.

I now head a SciComm group,
with my Honours near in view;
my brain no longer feels like soup,
this part of life feels true.

Just north of one Port Phillip Bay,
water and eyes an ocean hew;
a young man breathes today,
a curious life he'd still pursue.

# | 26 |

# Man's best friend

### JAKE LAMMI

As I lay on a large expanse of granite trying to warm up after an exceptionally cold swim in a glacier-fed alpine lake, I looked over at my coworker, Ranger, savouring the life-giving sunshine and cool mountain breeze. I couldn't help but reminisce on just how far the two of us had come together over the past three years.

At first glance, Ranger might seem just like any other coworker. He is supremely focused on getting work done, loves to hike long distances, and always brings a positive attitude to the "office." However, if you take a closer look, you might notice a few distinctions. He has a short nub-like tail, overly muscled stubby legs, and ears that bear a striking resemblance to those of Dobby the house-elf from *Harry Potter*. That's because Ranger, my coworker, is a wildlife detection dog with Rogue Detection Teams.

I graduated from college a few years back, with a degree in biology. I was just as lost as most after graduation, wondering what to do with my new degree. I knew I wanted to do something with conservation, and I knew I wanted to work outdoors, but so far, all I managed was a short-term sustainability job at a small university and a part-time track coaching job. I regularly checked wildlife job boards and talked with professors and colleagues who were professionals in the field of conservation. But in the end, it was an article in *Sierra* magazine that pointed me toward my next adventure: becoming a wildlife detection dog handler.

I had grown up always having at least two dogs in the house, but I had never thought about working with them until I read an article about an organisation that rescues high drive shelter dogs and gives them an outlet for their obsessive drive by teaching them to sniff for wildlife signs on conservation projects. I immediately emailed the program, asking for any information they could give me about how to get involved in this obscure field, and to my great surprise, someone emailed me back! While based in Washington, they just so happened to have a representative in my home state of Minnesota less than an hour from where I was living. We made an appointment to meet at a local park to "interview."

Minnesota in January is bitterly cold but despite that, I thought my meeting went well because I played fetch with two detection dogs. I had read that detection dogs were often obsessed with playing ball, as that is their reward when they located the correct target in the field. But I was nowhere near prepared for the level of obsession I saw during that short meeting. It was easily 10 degrees below zero with gusting wind and this dog looked like he would have happily played ball until his drool froze his mouth shut. I was hooked!

I applied for and was accepted into a training class that was being hosted in Washington State. It is difficult to hire for detection dog work as it takes a unique person to hike around for eight hours off-trail - solo - while working together with a ball-crazy shelter dog. They were inviting eight people to move to Washington for a month and would choose a subset to hire from those who excelled at the course. At first, it sounded a little like *The Hunger Games*, with the weakest member being cut from the team. I think as conservationists, we can all agree that this field is challenging to break into, as there are so many people who are interested, but not enough jobs or funding in conservation.

Turns out, training was different than anything I could have ever expected. One might think that it takes a long time to train a wildlife detection dog, but the dogs can be trained in a matter of weeks. It's the handler that takes several months and even a couple of years of instruction before being able to effectively communicate and work with a dog in the field. The reason being: just about every problem a person might be having with a dog can be traced back to poor communication between handler and dog, and many people do not accept this critique very stoically.

Dogs are adept at reading and sniffing human body language as well as subtle shifts in our moods. We are in constant communication with them whether we know it or not. Therefore, every exercise and training question we received in the class was focused on getting the trainees, to think from the dog's perspective. At first, this was quite a challenge. I failed more often than I succeeded and no matter how bad I wanted it, sheer force of will was not enough to change my old thought process. Dogs are not machines. It takes creating teamwork and a bond between a human and a dog to be able to work well together. It took a long time, but I eventually started to consistently think in this new way and was one of two people from my training class to be hired. That's when I met Ranger.

Ranger and I were first thrust together for a wolf scat study in northeast Washington. I thought I was prepared for the fieldwork after the class, but I soon learned that I still had a lot to learn. Ranger would consistently refuse to give the ball back, which slows down the field-work. He tried to trick me by alerting me to every single scat pile (data) in the hopes of getting to play with his ball, and he generally ran around like a chicken with his head cut off, becoming tired well before we had reached our daily survey objectives. He and I still had a lot of team building to do and in those first few surveys, I learned a lot about my-self, my perceptions, misconceptions, and eventually to trust this new bond Ranger and I were forming.

Ranger ended up being my best teacher.

Since those crazy first days, we've been inseparable. Over the next three years, Ranger and I would go on to work on 15 projects together and survey for over ten different species. We lived out of my rig, trav-elling from one project to the next. Sometimes it felt like a circus show; we were that busy. We climbed mountains, crawled through dense un-derbrush, forded rivers and post-holed through the snow together, just the two of us. We dealt with unpredictable weather, injury and extreme loneliness, but one thing we had never tackled together was a month-long backpacking trip.

This past summer, Ranger and I worked on a back-country study in Yosemite National Park. We were surveying for scat from mountain lion *(Puma concolor)* and the endangered Sierra Nevada red fox *(Vulpes vulpes necator)* and ended up having a five-week solo backpacking stint. I had done a few smaller back-country trips and thought I was prepared for the rigours of the back-country, but just like my first project in northeast Washington, I found I still had plenty to learn.

If you have never been to Yosemite National Park, one of the first things you will notice is that the terrain is incredibly steep and rocky. Part of doing detection dog work is being able to go where your target species live as well as explore other areas to learn whether or not they also travel through or utilise different landscapes. Both mountain lion and red fox tend to travel over high ridges and mountains which means, that's where Ranger and I had to get to.

I thought these mountain passes and knife-edge ridge-lines would be the most challenging rigours of back-country backpacking with a dog. In the end, what was most daunting was to be so completely isolated. I had no cell phone, which, while normal on our projects, we at least try to work as a team and come back after a day of surveys to a communal camping spot. In Yosemite, it was just me and Ranger, all alone, day after day. There was no fellow human to say 'hi' to at the end of a 15-mile survey. On some days it would hit me, looking around: it was just me, my canine coworker and hundreds of square miles of wilderness to search. However, as with many other difficult projects before this one, I had Ranger. Ranger's endless drive and goofy personality were what fueled me during four thousand-foot climbs and hair-raising descents down snowy glaciers. He was there for me when I got lonely and all too thrilled to share in my joy at finding our target species.

Stealing another glance at Ranger, now basking in the warmth of the mountain sun, I realise that although the life of a wildlife detection dog team can be lonely and difficult, none of us is ever truly alone as long as we have our canine companions alongside us. I feel more fortunate than most in our field precisely because I always have my canine counterpart with me to ward off the really lonely nights and quiet days.

# | 27 |

# Defined by our actions, not by their words

## CONNIE BAKER-ARNEY

*"You are not clever enough to be a vet..."*
*"You need to work harder."*
*"Are you sure you do not have undiagnosed dyscalculia?"*
*"I do not understand how you plan to achieve all of these dreams and plans."*
*"You need to get your head out of the clouds."*

These are some of the words I grew up with. These are the words that inspired me to prove everyone wrong.

I am a Kentish girl from England, I grew up in a home that was not particularly wealthy. We did not have holidays abroad or explore the outdoors at every opportunity. In fact, I spent a lot of time in the house as a child using my imagination to guide me. My entire life has been dictated by my want - if not need - to work with animals. It's like this uncontrollable drive in me: to protect and to save.

I, like so many others, had dreams of a veterinary career: my teenage years were dedicated to everything animal they could be. I worked on farms, in vets, in zoos, and boarding kennels. I took on part-time employment as soon as I could to fund these volunteer placements. I will never forget the feeling the first time I was accepted on a zoo placement. Finally feeling like I was a step closer to achieving my dream.

Veterinary school is difficult to get into, and I had always been aware of this. Always aware that I might not be good enough but also unsure that veterinary care was really going to be my passion in life. Alongside this, I was surrounded by people who did not believe in me as much as they should have done.

One teacher had no belief in me whatsoever. In an A-level chemistry lesson, where I already felt anxious and concerned that I was not good enough, my teacher blew up at me one drizzly Friday morning, apparently enraged by my "weakness". She stood in front of me and my entire class, shouting in my face about how I must be lazy, that I must have special educational needs, and that I would never get to university with my work ethic. My usual confidence dissipated, my resolve crumbled: it was immediate numbness. I will never be good enough and she confirmed that for me. But you know what? She was wrong.

That pinnacle moment in A-level chemistry has shaped me. It pushed me to quit sixth form and it set me on a new path; I went to college and studied animal management. There was a sense of freedom, a weight lifted. And why? Because I was another step closer to the goal. I was fighting against disbelief and the odds. This led me to my degree in animal behaviour.

I loved my degree: it fed my passions, it gave me confidence and it made me believe in myself. I was even lucky enough to get a job working in the animal unit there. It did mean sacrificing all my weekends for two years, but I didn't care. I had made it.

There was a moment in my car in that first week of work. I just sat there, glassy-eyed, staring out of my window. Covered in goosebumps and wanting to pinch myself. How had I done this? I had finally got into the industry: all those years of longing, wanting and working and I had done it. Minimum pay, zero-hour contract but to me, this was the dream. This was achieving.

I sacrificed a great deal during my undergrad to be successful and that meant my relationships and friendships were strained under the pressure of me constantly working day and night. Some of those people did not understand why I was so committed, and I am not sure they ever will. I was one of the lucky ones that had a job waiting for me at the end of my degree. Full-time work as a Technical Instructor managing animal sections and teaching students. An amazing job to cruise into after a degree.

And yet, I wasn't happy or fulfilled. Two years earlier, I had felt like I had made it with my zero-hour contract. Now, here I was staring glassy-eyed at my computer screen feeling like a bit of a cop-out.

One bad day later, I quit my job and booked a one-way ticket to Australia. I needed some time to level my head and decide what I want to do with my life. I could travel, I could learn about Australia's fauna and flora and maybe pick up some new skills along the way. I stayed in Australia for a year, travelled through every state, logged my wildlife spotting and learnt the behaviour of marsupials. Alongside this, I entered the racehorse industry. I believe in objectivity and I wanted to view this controversial industry from the inside. Plus, I was absolutely terrified of horses- what better way to combat a fear? I did combat that fear, in fact: I smashed it. I now love horses, I could write and write about how they changed my life, how they make my heart feel full- but maybe I'll save that for another time. Australia gave me that feeling of making it, above and beyond. I had taken another step closer to that goal of being an animal saviour.

This meant that leaving Australia and returning to England was incredibly difficult. I had spent a year moving forward, travelling, learning and discovering new things. Standing on boats and watching whales and dolphins, conversing with researchers, students and other tourists. Exploring islands meeting and seeing animals that you can only dream of.

Then suddenly, I was back in reality. I was static, going through the motions. My mental health utterly plummeted at this point. This was somewhat balanced out by meeting the love of my life and now fiancé, who showed me that there was more to the UK than grey squirrels. I applied for over 50 jobs, got four job interviews and three offers. I accepted a job in Animal Science lecturing, because I thought this was going to be the right move for me. I'm still not sure now whether it was.

Teaching students had some rewards, I cannot deny. However, there was no earth-shattering moments, goosebumps, excitement or pride that I had made it. To be truthful, I felt like a failure. I spent six months of my time working here going in complete circles about my static and miserable situation. I was put on anti-depressants to manage my anxiety attacks. I felt as though I had come so far, only to plummet so dramatically. Were they right about me not being good enough?

## Making a change

The only way out of this situation was to make a change. I applied for a master's degree in Northern Ireland and I made plans with my fiancé, Nick, to change how we viewed our lives whilst working statically. We bought a van, converted it into a camper and decided to commit our spare moments to wildlife, to videos, and to educating the young and the elderly. Sharing our passions day in and day out. Exploring in the van and searching for animals fills my heart with joy, it gives me that buzz. Every day and every time that I do this, I am learning and growing. It makes me buzz for my future career prospects.

That leads me to today where I now live in Northern Ireland, after working as a research assistant in a marine biology lab. During this time, I looked at microplastics and invasive species to complete my master's in animal behaviour and now, I am a PhD candidate.

I am working on a dream research project- studying fallow deer and researching maternal stress. Sometimes I pinch myself that I am working in a long-term behaviour study with such beautiful creatures. Reflecting on my present, I realise that all the rubbish sacrificial moments where I was convinced I was not enough or that I had failed, where I was told *"no"* by others, told to stop trying or to give up. I realise all those moments were worth it and all those people were wrong. I wanted a life with animals. I wanted to learn, to care and to protect them and *I am.* From animal care, to research, to saving an ant or a bee; all these things are positive contributions that we have all made in some way or another.

It is evident that we are not defined by the words or thoughts of others. We are defined only by our own actions.

# | 28 |

# The beauty of rock bottom

## NICHOLAS HORNE

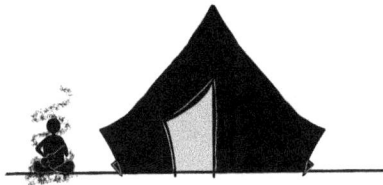

I had written a draft of this story a while ago, but now I really want to get these words out there because I think it could be helpful for many people. However, it was extremely difficult to write about my past experiences. So why did I reopen the document and start writing once more? Because I am struggling, and reflecting on when I was mentally at my worst may help me get some perspective on how far I have come. Hopefully, it may also help others to do something similar.

So, here is the time I hit rock bottom.

## A big step forward

I had just handed in my Masters thesis whilst studying Applied Marine and Fisheries Ecology at Aberdeen. That year of study had turned me from someone who liked science into a scientist - it was the toughest, but the best year of my life. Despite the pride I had in what I'd accomplished, I was still shocked and overjoyed when my thesis supervisor offered me a short-term contract working for one of his PhD students. Straight out of my Masters into working at my first paid job as a marine biologist. I got stuck straight in.

> *"We will sort out the contract soon as possible...No worries."*

So, there I was with three months of paid work ahead of me. To top it off, I had an interview for an amazing PhD up in the far north of Scotland, the dream. It was looking at collision risk between birds and onshore wind turbines. It involved three things I love: behavioural ecology, modelling and massive gigantic birds of prey! So, in between my days of hard work sorting through salmon scales, I got stuck into cramming as much info into myself ready for my interview to get my dream PhD. I was so happy.

After an amazing first week in the lab, I headed up to my interview and had a great couple of days. The interview went amazing and I felt like I couldn't have possibly done any better – it felt like things were meant to be. I headed back down to Aberdeen and continued with my work waiting to hear about the PhD. I was still waiting on that contract to get sorted but was sure that everything was going to be fine.

## The slope

After a few days, I found out that I didn't get the PhD position. It went to someone else – someone I knew. I was extremely happy for her, but that didn't remove the feeling of failure I had in my stomach. Sorting salmon scales suddenly went from being an organised lab project to a slow and monotonous process that my brain didn't want to engage in at all. Instead, my brain wanted to focus on every last detail of the interview, *"What did I do wrong? Will I ever find a PhD like that?"* But, I still had the job to get on with and I was sure, given my foot in the door, that more opportunities would arise.

After another week of sorting through scales, I had run out of money and my accommodation had run out. And still, no contract. I had to leave. In the space of a week, my dream PhD and my first job as a marine biologist had slipped through my fingers. *Failure.*

Did I even care enough about marine biology if I wouldn't work for free any longer?

## Tumbling to rock bottom

Back home living with my parents seemed like the ultimate kick in the teeth. I understand that I am an extremely lucky and privileged person. I love my parents to bits and they have always been there to take me in rent-free and support my dreams. None of what comes next is in any way caused by them.

Arriving back home, I had to find some work in a restaurant. Working in a place that shuts at 10 pm, in a city, with a bunch of 18 to 21-year-olds meant one thing: after-work drinks. I was spending 100% of my income on going out and drinking to numb the pain. That is all it did: numb it. My head would not let me forget the empty feeling I had going on. Things continued to get worse, aided by all the rejections I was getting from jobs and PhDs (I must have applied for nearly 50 positions at this time). Nothing seemed to make me happy.

The realisation of how bad my mental health had become, hit me like a brick one morning.

I had woken up, feeling awful as per usual. Managed to heave myself out of bed and started to make my way downstairs to get some food. Those ten paces to the top of the stairs seemed to be the hardest thing in the world at that moment. I reached the staircase, collapsed to the ground and began crying. It was like a flood of emotion. Taking over my whole body my heart pounding, my legs giving way and I had lost all control.

This was my rock bottom.

Yes, things could get worse, but I decided that day that I wouldn't let it. I realised that I was the only person that could truly solve my problems. So first of all, I made a drastic change and stopped spending 100% of my income on alcohol and dropped it to about 50% (I had to have some fun!). This meant I could begin saving money. I wasn't sure yet what for, but I knew I needed money for when the opportunity arose.

## *A small step in the right direction.*

I began searching harder than before. More applications and more searching for what to do next. I looked into visas for New Zealand and Canada and all the opportunities that could arise. I did find a voluntary position as the digital editor of a conservation-based publication which had me on the board of a charity and running a website. It took up very little of my time, but it kept me thinking about conservation and gave me something for my brain to chew on.

I continued with the search and finally an opportunity worth taking the leap for came up. I had saved enough money to put myself forward for an unpaid position monitoring marine mammals in Ireland. Yes, *unpaid.* However, I decided this was a great opportunity and worth it because, well: I was pretty desperate. But also, the following adult reasons:

- There was no fee and accommodation was included.
- It was with University College, Cork and therefore a great opportunity to network for future prospects
- Ireland is awesome (maybe not so adult but still)

The three months I had there were fantastic and funnily enough, didn't end. These situations are where the phrase '*You make your own luck*' comes from. A job looking at seals stealing fish from fishing lines with the same people suddenly became available due to someone else not being able to do it. They needed someone that knew the area, had experience and ideally, someone that knew those staying in the accommodation that the research team was in. Right place, right time. I immediately put myself forward for this job and got it. I had a contract posted over in no time. A CONTRACT! My first paid job as a marine biologist was here. I was over the moon.

It doesn't end there. Being on the same island as a close friend – he visited me out in County Mayo. Upon his return to the Queen's University, Belfast Marine Laboratory he spoke to his supervisor about his weekend visiting me, and my work on marine mammals. Funnily enough, they were working on a PhD proposal looking at seals and collision risk and asked if it would interest me. Obviously, I was keen as can be! So here I am, at my desk doing my PhD on seals and collision risk.

## The lesson

Am I still struggling with the lows? 100% but I do know that when I think back to that moment on my parent's staircase, crying, I am in a much better place than I was then. The biggest lesson I learnt from this situation is that surrounding yourself with people that understand what you are going through is unbelievably helpful!

My friends and family back in Essex are great but they do not know much about conservation work or academia, so talking to them about my problems wasn't very helpful. But, once I had surrounded myself with like-minded people, my confidence and drive started to return. I had others surrounding me who were in the same situation, and it felt great to share the burden but also to be there to support them.

I am so grateful now when I look around me, at my fiancé – who is also a scientist, gives me so much support and is always there to talk to. I am also grateful for my network of friends in science that have been also through the same situations as I have, and they are so open and ready to talk about my worries at a drop of a hat. So, a support network where you can go to talk and also go to support others is invaluable.

# | 29 |

# What no one tells you about careers in wildlife

## MADHUSHRI MUDKE

Ten years ago, the dilemma of finding an appropriate career in wildlife that allowed me to work with wildlife was a daunting path to take. I began to ruminate – 'I love animals, I love being in the wild' and 'we must save wildlife and forests to combat climate change'. Back in 2010, climate change was haunting humans like it is today and biodiversity losses were still alarming. Despite the popularity of National Geographic and Animal Planet, choosing a career that would last a lifetime and allow you to work with wildlife is more difficult than performing a pole vault. The thing with careers in wildlife is that they aren't as straightforward as other professional careers like law, engineering or medical sciences. After graduating from a generic professional college, most people end up getting jobs at established organisations. Such jobs not only provide financial security but also self-validation, societal respect and a sense of community through conferences, work meetings and events.

## Focus on your strengths to pursue a wildlife career

Let's start by acknowledging that it is extremely difficult to understand yourself, let alone understand professional working conditions within the wildlife sector. It is important to look beyond the pleasures that being in the wild or being with animals can offer. A career in wildlife is diverse and knowing your strengths is important for a great start. So your first step should be to note down all your strengths: personal and professional. Imagine you have a vast canvas in front of you – what's the most prominent colour, rather, strength, that you want to see in there? Your greatest strength is your biggest skill that allows you to reach a larger goal. For example: spoken and written language, filming, coding, social media, creativity, drawing, writing or a successfully completed degree. All of these could be enlisted as your strengths. Now work out a path that begins with your own strengths and encompasses wildlife within it.

For me, working in academia is fulfilling and fun. However, no one told me about the challenges in academia for a woman researcher in a developing country. Through all these years, I learnt that an important skill is to navigate through the difficulties and look beyond academia and at larger interests and problems that require you to continue working on what academia prompts you to.

For example – writing a small academic paper can help you in writing a big and important one down the line.

When I was choosing research as a career and looking for opportunities to pursue a PhD, I thought that generating new information scientifically is the only way forward. Since I love reading, writing and analysing data, my skills and strengths have allowed me to navigate through the challenges of scientific careers. Recently, I have come to realise that scientific careers are not for everyone and that pursuing one can be mentally, physically and financially extremely exhausting! Therefore, when anyone asks me how to 'work with wildlife', I start by stating that there are too many subfields and facets to wildlife. Along with that, just like me, several people believe that they are doing the right thing. They'll tell you that there's only one right way of working with wildlife, but trust me there are several varied paths.

## Thinking beyond existing wildlife career options

Scientific research, professional filming, photography or even working with established NGOs/organisations as project managers or leads can set one on a successful wildlife career path. Despite these well-known options, trying out teaching, entrepreneurship, writing, art and content creation might also allow you to 'work with wildlife'. A professional degree and building interpersonal relationships will surely help a newcomer. But these established paths might seem daunting due to the inability to maintain a proper work-life balance, difficult working conditions during fieldwork, financial security and added mental health issues. The lack of a support system and internalised misogyny has been highlighted as one of the major reasons why women tend to fall out from these well-known career paths. Given all of this, leaving out established career paths and venturing into the newer ones might be ideal.

There's ample scope for independent entrepreneurs who are now coming up with ways to work with biodiversity and material goods. Technology is being used extensively to create space where wildlife could benefit from capitalism. Have you heard of biodiversity-friendly coffee? Apart from that, there is also a rise in grant programs for creative writing and art. This may allow creators to travel to forests and learn about animals while pursuing art. Such projects will also put food on the table. A career in wildlife could mean anything from running a charity that supports wildlife to working at a zoo. However, one must start by appreciating various *unexplored* nuances to careers in wildlife.

## The constant lack of self-worth

Most people seeking to find their wildlife career paths end up feeling dejected, lonely, financially unstable and riddled with imposter syndrome at some point in their lives. This is mainly because wildlife and environmental concerns are largely value and convenience-driven. Each person ends up setting his/her own goal while trying to fulfil a larger goal of 'working with wildlife' or 'saving the forests' at one's level. For instance, if I am making a film on red pandas, I am helping the cause of saving the red pandas. But a film may not directly stop a poacher or a large infrastructure project that encroached on the red panda habitat. There's a missing link here that one needs to understand.

Additionally, putting environmental and wildlife concerns before your own individual self can cause mental exhaustion. The lifestyle choices one prefers can add to the constant lack of self-worth and de-motivation. Questions like: are my daily living habits sustainable or are they further endangering biodiversity and forests, should I buy groceries packaged in plastic that threaten more than 250 species of animals or shall I find ways to reduce my plastic waste? Living with such thoughts can lead to agitation and disquiet. Entangled thoughts at the back of the mind of an environmentally conscious person will make things worse even before one chooses an appropriate career. However, let me tell you that this is a great start in knowing who you really are. Therefore, choosing an appropriate career in wildlife that'll also allow you to lead (a somewhat) environmentally conscious life might actually be the right thing to do.

Wildlife and environmental crises are too large to tackle at an individual level. This, I think, is the number one reason behind a lack of a sense of community or togetherness among people who are pursuing wildlife careers. There's no right or wrong, no single goal and no single answer. Everyone involved has set their own separate goals via different approaches that are convenient to them. People working in wildlife sectors have different worldviews, opinions and methods available to keep the 'living with wildlife' or 'I am saving the forests' spirit alive.

# | 30 |

# Love and science

## JADE ARNESON

I grew up fortunate enough to have woods, a creek, and 40 acres of farmland as my back yard. I spent considerable time outside as a kid. I clearly remember walking through the field and standing atop the hay bales, wandering down to the creek to have a picnic lunch, and staring up at that big old willow tree wondering if raccoons or an owl had made a home in the cavity left behind by a fallen limb. I also loved animals and remember being very fond of my cats. Often I'd pretend I was a veterinarian – my cats or my numerous Beanie Babies were my patients.

Growing up, I would always say I wanted to be a veterinarian or a scientist. What did it mean to be a scientist though? I remember envisioning flasks, titrations, white coats and eye protection as a young kid. Of course, now I know that science is very broad and inclusive of many disciplines, like conservation and even veterinary medicine, but as a young kid, I had no idea. I can also say that being a 'scientist' was not something my teachers or parents ever unpacked for me; it also wasn't a career path that I was encouraged to pursue. Veterinary medicine, however, was a career I knew a little bit more about: one that seemed easier met and more clearly defined, and also one that my teachers and parents supported.

Now that I'm 25 years old, with a Bachelor of Science and a Master of Science in the works, which career path did I end up pursuing? I chose conservation, the perfect intersection between my dreams as a kid to be a veterinarian and a scientist. I gave veterinary medicine a try for quite some time, shadowing small animal veterinarians in high school and college and working on a dairy farm with their large animal vet. I was pre-vet through junior year of college, with a wildlife ecology major – I planned to be a wildlife veterinarian. One day though, while sitting in organic chemistry, a pre-vet requirement, I decided I didn't want to be a veterinarian, and not long after, I decided I didn't want to be in organic chemistry, so I dropped it. I can't fully explain my reasons to no longer pursue veterinary medicine, but I can say that at the time, I was very excited and motivated by the thought that I'd be a wildlife biologist.

That excitement I felt as a junior in college at the thought of being a wildlife biologist is something that has somewhat waned today. It pains me to accept that and to think about why that is, but thanks to this great group, I can spend some time thinking and writing about it and hopefully find people who feel the same way.

Before graduating with my Bachelor of Science in Wildlife Ecology, I had a job lined up with a non-profit conservation organization. I remember thinking how great that was, to not even have graduated yet and have a job lined up; some of my classmates couldn't say the same though. Before I knew it, my summer employment with the organization was coming to a close and I was spending every evening in front of my computer searching the job boards. The thought of the next job was exciting but also equally terrifying and wrought with anxiety. Even now, I remember that MS Word document on my computer of jobs I had found and wanted to apply for: jobs that were all across the country and several of which involved work on great grey owls and California spotted owls. I never applied to any of those jobs though. Instead, I emailed the local federal conservation office and asked if there were any temporary employment or volunteer opportunities. They wrote back indicating they had volunteer opportunities, so I arranged my first day to be there to volunteer, meanwhile calling back my old boss at the restaurant to ask if she needed another staff member and how many hours a week I could get bartending. Thankfully, she needed the help and then the reality hit that after four years of pursuing a degree in conservation, I'd be relying on the restaurant for income.

Rewind to January 2015, four months before my graduation, I met a guy when I was back home for the weekend; he is now my fiancé. On the 8th of March 2015, he asked me out while driving back home after a trip to the Apostle Islands National Lakeshore to see the ice caves. I will never forget hesitating to respond to him and prefacing my answer with,

*"I just want you to know that this field isn't very stable and I want to pursue it whole-heartedly. I'm just getting started"*

and he said,

*"I know, I'll support you."*

That day changed the outlook of my career. Thankfully my position with the non-profit conservation organization was only one hour away from home so he came up every weekend to visit or I came back home to visit. The reality of the situation, and of how torn I'd end up feeling, didn't come until I decided to never apply to those jobs out west working with great grey owls and California spotted owls after my time with the non-profit organization was up because that would mean leaving him. Leaving him for a field season was something I knew I could handle, but could he? And how would I tell him? How would that conversation go? Would he be supportive like he promised? At least I had the restaurant.

The day came for me to start volunteering at the federal conservation office – I'd be assisting the fisheries department. I volunteered for two days and much to my amazement, I was offered a full-time temporary position as a Biological Science Technician, which would supplement my income from the restaurant. *"This never happens"*, I kept telling myself, and the timing couldn't have been better. It meant that instead of sharing my concerns with Branden, I could share my happiness. It meant that he and I could continue to live near one another and not have my career come between us. *"We should be able to keep you on for a few months"* is what I was told and that was a huge relief for me. I ended up really enjoying my time at the office and felt satisfied that I was continuing to grow as a professional versus just sitting behind the bar of a restaurant all day.

But then the time came again to look for jobs and this time, I applied for every position that sparked my interest, no matter where in the country it was. One day, I collected myself and told Branden the jobs I had applied to and that I felt strongly about several and it ended up resulting in tears, heartache, and confusion.

I don't know how I ended up convincing him: maybe he just came around and remembered the words we exchanged when he asked me out, but I took one of the jobs I had applied to and funny enough, it was a job working with California spotted owls – it's like I was being gifted another chance, perhaps one with better timing. We travelled out to California together by car, making a week-long trip out of it. Seeing him off on the plane back home wasn't easy for either of us, but we got through it. That job was what I needed. I got to go off on a crazy wildlife adventure, the one you always hope for and dream about. I was also able to dedicate time to myself and learn more about who I am. On top of all of that, I met some great people, saw the beauty that is California, and helped conserve a species of owl that has seen significant population declines and habitat loss. While I was out there, Branden and my best friend visited me. Branden's mom also met me out there to drive home with me, so I got to connect with her better, and when I got home, Branden proposed to me – this time my hesitation was all in asking if he had asked for my dad's permission first!

Branden and I have been together for almost four years now. After my job working with California spotted owls, I saw myself back at the restaurant behind the bar but I also found myself back at the federal conservation office working full time. And thanks to a former co-worker at the office, I now am a graduate student and a research assistant at a local university pursuing a degree in environmental science and policy and working on restoring wild rice to coastal wetlands in the Bay of Green Bay, Lake Michigan, USA. Branden and I still live together: we moved about 40 minutes north so I could be closer to the university. In doing so we bought a house and now have also added a German Shepherd and two cats into the mix. I often wonder how long we'll reside here and what lies ahead of me once I graduate with my Master of Science degree. Part of me worries that there will be more tough conversations to be had and tears to be shed dependent on where I find job opportunities or get job offers, but I have to trust that Branden will support whatever decision I make.

Now that I've shared my story and will bring it to a close, I want to end by saying that I've joined this group because I still struggle with forging my way through this fantastically challenging and sometimes defeating career. Sometimes I wonder if it's too soon to have bought a house, to have signed on to dedicating my time and energy and resources to two cats and a dog, to have gotten engaged, even to have accepted this graduate position. I even sometimes doubt my place within this field and question if I should continue pursuing it and I think a lot of that doubt stems from the fact that this field is very unstable, competitive, and financially burdensome. Doubt also stems from friends and family members, and sometimes even significant others, who struggle to understand your passions and tenacity for this career.

In addition, doubt is contributed by members of the public, the stake-holders that many of us in conservation work for, and politics.

There is no doubt though that so far I've been blessed to have enough opportunities locally to keep me busy, and what's better than getting to learn more about the landscape that has been a large part of my life thus far and contribute to conserving it?

# | 31 |

# Finding your voice and following dreams

## JESSICA PINDER

*Growing up Green*

It could be easily said that I became a conservationist because much of my childhood was spent outdoors. Almost all my family holidays involved swimming in crystal clear rivers, hiking through bushland heavy with the creaks of cicadas or cross-country skiing across Victoria's snowy mountains. My parents even renamed the iconic Blundstone boot as *Jessie Boots* because I wore them with everything, even with the terrible pink and frilly dresses that I was on occasion forced into.

For as long as I remember, I have always known myself to be an inquisitive tomboy, adventurer and self-professed nature nerd. I mention this because I think early childhood is a completely fascinating and influential period in our lives. I'm so obsessed with it that I completed my master's thesis and published my first paper on the topic of how childhood experiences play a formative influence on our conservation attitudes and career choices. It makes a lot of sense to think that as nature lovers, our earliest experiences in the wild must have greatly influenced our life trajectory. But curiously, my study found that conservationists didn't actually spend any more time outdoors than other kids! As it turned out, our childhood experiences in nature weren't very different from those of aspiring lawyers, teachers, and artists.

It's highly likely that the experiences that have led us to care about conservation are far more complex than you might first think. Caring about conservation is a complex socialisation progress, one that is value-based and primarily influenced by the people around us rather than our experiences of nature on their own. This doesn't mean that experiencing nature as kids isn't important, but to enhance conservation concerns, these nature experiences must be combined with conservation values and knowledge to make the lesson stick.

The best part is that these conservation messages can come from a great variety of sources. For example, kids who became conservationists frequently picked up messages from the stories they loved most in childhood, their favourite teachers and the *biospheric* or nature centred values they were taught from their parents, social groups, and involvement in environmental organisations. In short, the secret to creating conservationists is through a person's social upbringing- by encouraging children to love and be kind to nature, to respect their role as custodians of our planet and to understand and care about environmental problems.

## Starting my conservation journey

Throughout my childhood, I remember soaking up every snippet of information about the natural world. From very early on in life I recall feeling a strong sense of purpose in championing threatened species and working to protect Australia's incredible natural landscapes. My first attempt at saving nature was to raise money for a local wildlife shelter, by selling photos of birds that I had taken with my dad, then a photojournalist. We spent a lot of time in nature reserves hiding out for fairy-wrens, stalking unsuspecting rose robins and kookaburras that cackled from treetops. I loved the patience it required: the waiting and watching the world around you until some beautiful animal crawled, hopped, ambled, or flew by.

Only a small amount of money was raised throughout the entire fundraising endeavour, primarily from sympathetic family members and members of the public bemused by an eight-year-old's strange passion. However, it was enough to donate to a wildlife shelter, which I visited and had the pleasure of learning about the stories of rescued kangaroo joeys, possums, and a cockatoo. In the years since, I tried many other initiatives, but in the absence of support or mentorship from a local environmental group, my impact was in reality very small.

As a young person who cared deeply about wildlife, I remember being very angry and disbelieving when I learned about the environmental damages that humans were causing to our planet. Throughout my teenage years, I felt like I was paralysed by feelings of powerlessness. As a naturally very shy and quiet person, discouraged by years of effort and disappointed expectations, I felt convinced that my voice was simply too small to be heard. I developed a very keen sense of anxiety for the future of our planet, which increased every year as I tried and failed to make an impact on the world around me. I put so much pressure on myself to make a difference that burnout, fatigue, and frustration soon became my norm.

I'm 27 years old now, and some days I still feel that same sense of anxiety tripping me up. At times, I am exhausted by the ever-growing list of actions: be plant-based, be zero waste, be carbon neutral, buy second-hand, make yourself, plant trees, conserve water, educate others, reduce energy use, volunteer, donate money, compost, recycle... How we need to act to save our planet go on forever, and my guilt at failing to achieve in every one of these areas grows.

For me, the problem with caring so much is that I want to do every-thing within my power, and the sheer volume of responsibility I take on leads me to inevitably fail. Sometimes these failures make me feel like a hypocrite, and I question the times in my life when I have spent over 40 hours a week working in a job that *isn't quite conservation.* I've drastically over-compensated these *failings* by volunteering far too much to make up for it and I inevitably ended up fatigued, disheartened and unable to make a difference. If you're keen to learn from my mis-takes here, know that finding a balance that works for you is essential. Taking action is of course an important part of healing ourselves and the planet but making sure your needs are taken care of first is very, very important.

## Forging my path

Growing up, as I did, in a country town in rural Victoria, I never felt there were many options for me career-wise. Environmentalism was pretty well frowned upon and when I decided to move to Melbourne to study ecology, some family members were quite vocal about their con-cerns that I was throwing my life away. I recall having numerous argu-ments about the time and money I would be wasting and feeling worn down by the countless cautions about how I'd never get a job. Everyone I knew said it was going to be hard, and they were right. But I can hon-estly say that I have no regrets. None whatsoever.

I think that no matter what path you choose to go down in life, there will always be challenges and risks. But if you want to give yourself the best chance at happiness, it's so important to stay true to your path. If you really want something, you have to be prepared to work hard for it, and, most importantly, never give up.

Since making that first big decision to leave my hometown, I've sup-ported myself through two degrees and volunteered my time for count-less organisations and research projects around Australia. I worked throughout all my studies in casual jobs, first at a bakery and later at a hiking store. Those casual jobs taught me some of the most valuable skills I've ever learnt, like how to be a good communicator, put a voice to my thoughts, and especially how to get over my fear of social interactions.

I landed my first job working with wildlife when I was 22. I was recruited as a wildlife presenter, and my job was educating young people around Melbourne about a diverse array of native Australian wildlife which visited the classrooms with me. I loved working so closely with the animals whose protection I cherished and did my utmost to inspire other young people to respect our planet and take action in their own lives and community. My public speaking improved tenfold, and I became confident presenting to larger groups. I grew so much as a person in that role and learned a lot about my own values and ethics surrounding wildlife tourism.

Around that time, I started volunteering for a youth-focused environmental NGO, as a member of their national youth leadership program. As this was my first experience as a member of an environmental organisation, I remember feeling so relieved to meet other passionate and like-minded young people who, like me, were doing their best to make a difference. I felt inspired to get back into research and took a big leap by moving my life to Queensland to study for a master's degree in conservation science.

Skipping ahead through 18 months of anxiety-fuelled study and other life dramas, I graduated from my master's with the most amazing group of young conservationists from all corners of the globe. I took over management of the program and worked to transform it into a tailored professional development experience for young people wanting to make a difference in the world.

And then something magical and unexpected happened.

I found out that I had been selected to attend Jane Goodall's 2019 Global Leadership Gathering at Windsor Castle UK, as Australia's youth delegate. When I found out, I could scarcely believe that I had earned this once-in-a-lifetime opportunity, simply by following through with my passion with hard work and determination. At this event, I sat next to Dr Jane Goodall over dinners and caused fits of laugher when I once accidentally spilled red wine on her when pouring. I had the opportunity to ask her incredible questions, and we had the most insightful conversations about the future, and how we can work together to give young people the best tools to create change.

But perhaps best of all from this experience was meeting young changemakers from all around the world, which expanded my mind hugely to the interconnectedness of the problems facing our planet. I am so proud that at this event, I was part of a team that ideated (and has since developed) a global habitat restoration campaign and resource for young people that will be launched in 2022. It's all about falling in love with the natural world around you and engaging in creative projects that help people and the planet- and it is aptly called 'Embrace the Wild'.

This conference also gave rise to another exciting opportunity in my life, as I was miraculously interviewed and featured in a National Geographic documentary being filmed called *Jane Goodall: The Hope*. In one scene, a surprise guest Harry, the Duke of Sussex (Prince Harry at the time) arrives at our conference and shares an adorable chimpanzee greeting with Dr Jane. He and I chat about the loss of connection between people and our planet and Harry raises his hand to offer me a high-five, which I, in true fashion, accept with characteristic awkwardness. The camera pans across to the hero of my early adult life – Dr Jane Goodall - who explains how young people, like me, like you, dear reader, are her greatest reasons for hope in a dark world. This sequence of events, immortalised on camera, is easily one of the prouder moments of my life. It is the greatest proof I have that even after years of thinking otherwise, my voice and what I have to say really does matter after all.

## Update: 2021

As it turns out, I never did get a wildlife conservation job straight after graduating from my studies as I had hoped. Instead, I joined a state government graduate program to learn more about how policy is made and developed. I worked on important challenges including water policy, First Nations social policy, and even tackled reform of our energy systems. While my day-to-day work during this time didn't involve interacting with wildlife or being out in nature, I am still so grateful to have had experiences in which I was continuously learning and delivering tangible positive impacts into the world.

More importantly, these roles also helped me to shift my focus from pure conservation to broader social change. I am now what I consider to be a community conservationist, and my passion lies in the intersection of nature-based solutions with the empowerment and self-determination of diverse, local communities. This people-centric approach to conservation happily means that I no longer feel the need to hate people as a rule, and I am confident that encouraging community leadership in the gradual transformation of our social and economic structures will create far more sustainable than other aggressive conservation strategies that be guilty of excluding, opposing, or blaming disadvantaged people. These new ways of thinking have opened so many doors for me, and I think have been critical to me finally breaking into the conservation policy field. I am so excited that I now get to spend my days helping regulate environmental protections. After so much time spent growing, learning, and building up my experience in other areas, I can confidently say that I am finally thriving in this role.

It has taken me far more time than I could have imagined getting here, but looking back, I am so glad that I took the courage to follow my dreams. They have been a guiding light through my experiences of complex trauma, financial hardship and all the other difficulties life has thrown at me, and they clearly defined me as the person I am now today: a passionate, community-minded conservationist who is proudly using her voice and ensuring so many other voices are heard along our journey toward a better world.

# | 32 |

# Finding my sparkle of joy again

## PRISCILLIA MIARD

Hi everyone, my name is Priscillia and I guess like many of you I have multiple unpaid jobs if I can call it like that. I am a researcher, a primatologist, a project manager, a photographer and many more.

I grew up not really knowing what I wanted to do and made it into university knowing I like nature and biology, but that was pretty much all. The school system puts you in boxes and for someone like me, who has a lot of imagination, it did not work well. I guess I was lucky to grow up in the countryside of France surrounded by farms, forests and animals. My love for nature started with my grandma who was a gardener for castles. It sounds pretty fancy but in reality, it was just amazing to spend time with her in the woods surrounding those beautiful houses. We always went to watch the deer coming and many other wonderful animals.

I am not sure where to start but the journey I am on has been amazing. Not always full of joy but also a lot of crying, sadness and along the way dealing with personal life to choose my own path. It took a bit of time for most of my family and friends to understand what I was doing.

I have now been working in the field of research and conservation on and off for the past eight years. It all started with an amazing internship in Borneo that I will always remember. For me, it was like being in one of those documentaries I always watched on TV since I was young. The jungle has a special feeling that I can't forget, and I definitely cannot stay out of it for too long. It is when I came back home from this that I realised something was missing. It hit me so hard, I just had to give up all my life there to go back. This meant leaving behind everything I thought I knew: friends, boyfriend and family. And so far, I have met amazing people along the way and made friends from all over the world. I travelled more than I ever thought I would when I grew up as my parents were not the explorer type at all. And going back was maybe the best decision I have ever made so far. I was feeling great, loving what I was doing and being at my best every day.

But recently I realised that doing this kind of work, even if amazing, is also mentally hard. It is a lonely path for many even if we are surrounded by people. Keeping mentally healthy and dealing with a lot is never easy and like everyone else, I had my fair share of issues. Happiness and keeping positive is what helped me go through all those years but recently I realised that I lost that sparkle of joy in me. I have been, for most of my life, the girl who is always happy and smiling, making myself and others feel better every day in a world that is not that easy. This was keeping me happy and motivated until it did not work out anymore. Maybe it is the path for most PhD students and everything that goes with it, I guess.

I have never felt so alone than in the past year even if I did not want to admit it. Of course, I love what I am doing, and I have some amazing friends and people who support me and help me when I am feeling down. But I have been working way too much, almost not taking a day off except when I was sick, and this for a non-negligible number of months. I think I was trying to keep focus or trying to run away from reality; who knows? I have been doing my PhD for the past two years and also running a project at the same time. So, this meant doing research but also managing and training volunteers and interns at the same time. I really enjoy it - don't take me wrong - but I always have a hard time focusing and always want to do more.

Then I went back home for a month to see my family and some people I had not seen for at least two years (some for more than 10 years). This is when everything hit me pretty hard, the return to reality if I can say like that. The realisation that I cannot keep on going like this and the anxiety that goes with it. For one month, I completely disconnected from my work and pretended as if I was like everyone else. Working a part-time job during that time to save money and meeting with people I had not seen in a long time. Being in an industry where it is hard to get a real paying job and a decent social life made me become pretty bad with myself and stressing too much about it as I really don't want to take a job I don't like. And of course, that pretty bad impostor syndrome we all have even if we are all doing amazing things every day.

When I came back to my home, in Malaysia, I decided to take more care of myself because I have been through really bad anxiety once and I told myself never again. I reconnected with many friends and was trying to decide what was the next step for me. All this was not as easy as I thought it would be when I realised it. I battled with a lot of anxiety regarding my personal life for a few months and the future that goes with it; not enough sleep, the stress of thesis writing and also problems of not eating properly and losing more than 7 kilos (5 in less than a month).

I have never been someone who needed others' approval to do what I think is right for myself. But I guess we all at some point have a hard time focusing on what is important. But all the hard work I am currently doing on myself during this hard time, reflecting on the way I was towards myself and others really paid off. Meditation, reading, listening to many great podcasts about mental health and a lot of outside activities were the key for me. I am still trying to get out of this negativity zone but once in a while it all comes back, but I will eventually get there. It might not be the same for everyone, but we just have to listen to our mind and body, what they are telling us and how we can move forward.

All I can say is that the amazing sparkle I had lost is finally getting back slowly and ready to bring joy and inspire others to do the same. Working in conservation is all about bringing joy and empathy to a world we love and want others to do the same.

It is not easy for me to talk and write about all this, but it is a step towards feeling better, realising that something is wrong, admitting it and working on it. And I know that I am not alone in this; we all have doubts, personal issues, relationship problems and other things messing up with our minds once in a while. I always got those: when will you get a real job? When will you get paid for what you are doing? Are you going to get married one day?

We live in a busy world and slowing down is never easy. Our brain is also wired to remember and work more on the negative experiences in our life than the positive ones. We also work in a field where we think we can always do more, we have to do more. But the most important thing is to remember why we are doing it: to bring change and happiness. And this cannot be done if we are not feeling alright ourselves.

So now after eight years in this field working mostly for free, even if I am lucky to have a scholarship and great sponsors at the moment, I think I am ready for a slightly different path. I think I need to get out of my comfort zone again and try on new challenges to keep that sparkle of joy that is keeping me going and continue to bring changes for a better world. I will try my best to succeed at those even if I know there will be failures and hard times along the way. But there will also be amazing moments and those are the most important.

Failures and hard times are what makes us move forward. They help us remember what is important and remind us to never give up.

# | 33 |

# This is my voice, a black voice, and I'm not sorry about it

### JAMES LEE

It is a time of unease in America with regard to the treatment of racial minorities.

Whether it be a young jogger shot like a dog in the street in Georgia, or a man having his neck crushed by a police officer in Minneapolis, or a birder having the cops called on him after simply politely asking someone to put their dog on a leash, these recent events have laid bare the vast disparity in the amount of fairness and justice afforded to black people and other people of colour (POCs) (or in their case, the lack thereof) that white Americans have long taken for granted.

It is an unfortunate truth that racism has been an ever-present force in our country, upheld by institutions designed to maintain the status quo and refuse to change. There is a double standard in our country where minorities are both judged with much stronger scrutiny and far harsher punishment than white people, and this isn't some phenomenon isolated to a few highly publicized incidents. This is the norm.

This is particularly true for POCs who seek to use the many incredible natural areas across the country. Speaking from personal experience as a black man in a predominantly white career field and often working in overwhelmingly white areas, I have long envied the sense of security and ease that my white colleagues must feel when they venture outdoors. To work and dwell in outdoor spaces without fear of judgement or baseless retribution, and to rejoice in studying and celebrating public lands and natural areas must be quite wonderful.

Do you think my fears are unfounded? What if I told you that I have been personally threatened with firearms for trespassing, even when I was not? Or if I was politely explaining the fieldwork I was performing to a suspicious landowner who took a photo of me without asking when she suspected me of some sort of wrongdoing? Because both of those things have happened to me, and instead of letting myself be angry at them for their unkindness towards me, I instead have to be grateful that things didn't escalate further in either such situation, and be thankful that I simply still have my life and health.

And it isn't even always direct interactions: in fact, much of the hostility POCs face in America is passive and atmospheric. Even during my last field job in northern Georgia, I can recall peacefully driving through that countryside and marvelling at the beautiful scenery of Southern Appalachia, only to have my reverie broken by the sight of a Confederate flag. Or the many instances where I've politely waved and smiled at people I've come across who instead ignored me at best and glared at me with contempt at worst, as if I were not welcome in those places.

This is not to say that I have never had positive interactions with non-POCs while performing fieldwork or recreating outdoors- quite the opposite. I've had many. But there is no guarantee that I or other POCs won't be confronted with hostility or violence when we go outdoors. Understanding this fear is part of the reason participation in outdoor recreation is dramatically lower among black people and other minorities than it is for white people.

And while we're at it, let's address why a lot of these areas are so overwhelmingly white in the first place. Why are there so few black farmers? What happened to the descendants of the black and Hispanic cowboys and homesteaders that once settled across the rural areas of America? Or the descendants of the Chinese immigrants who toiled in mines and built railroads that stretched across the west? Or the myriad of indigenous peoples who inhabited this continent for thousands of years? Once they lost their utility to white settlers, they were hurried away with violence, and now are assumed to be "ignorant city folk" by many of those still living in rural areas (or in the case of indigenous peoples, cruelly stripped of their ancestral homelands and placed on reservations). And that is why so many POCs feel unwelcome in such areas.

A lot of people are saying things like *"We need to have a conversation"* or are asking *"What are we gonna do?"* as if the answer to these issues we're facing are arbitrary and must be determined by some great convention. To me, they're fairly simple and apparent: we need to dramatically adjust social systems and institutions to be fairer for everyone, and dismantle the institutions that cannot be changed.

We need to recognize how historical oppression and biases are still strong and pervasive to this day and make active efforts to combat them both through legislation and cultural change. We need to hold police officers to a much higher standard and hold them accountable to the communities they're sworn to protect so that they don't abuse them.

To the naysayers who might say *"If you don't like it, then leave"*, I say this much: no. I have as much a right to work and dwell in these places as you, and while it may not be as safe or welcoming for me to be doing so, going outdoors is what I love to do. I grew up spending countless hours in the woods near my home, which was a massive influence on why I became interested in pursuing wildlife as a career in the first place. Everyone in our country should have the same luxury of going outdoors and utilizing public lands and not being afraid of other people in doing so.

This is my voice, a black voice, and I'm not sorry about it.

# | 34 |

# Becoming visible

## MATTHEW LEFOE

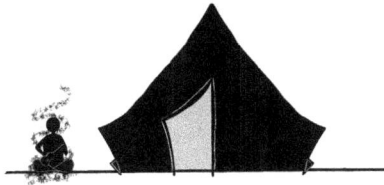

In this industry, we are conditioned to be constantly networking/ up-skilling/career-building to be competitive for jobs. Our job pool has always been small, and that's why we feel a constant need to make ourselves the most employable candidate. Those jobs have since dwindled further in response to the economic effect of this global pandemic. But do I regret choosing this career path though? Absolutely not. If I didn't embark on this journey, I wouldn't have been respecting who I truly am. So let's see if putting pen to paper (or finger to keyboard) about my conservation journey can provide the therapeutic outlet that I'm seeking.

I grew up in Bobinawarrah, Victoria. To call it a town is a bit of a stretch: it's an area with a dozen or so farming families living in it. From our veranda, you could see Mount Buffalo and in winter, we could see when it was capped with snow. We were on the cusp of the floodplains: a vast network of rivers and creeks dotted with ancient river red gums and strongly perfumed silver wattle. Dry open eucalypt woodlands dominated the networks of foothills skirting our property. I spent entire school holidays/weekends exploring this diversity of habitats, both alone and with my siblings. I'd spend hours yabbying (trying to catch yabbies, a type of freshwater crayfish native to Australia) in the creeks, scouting for wedge-tailed eagle nests, and trying to track goannas (monitor lizards) through the forest. This was my home and my escape, where I felt free to be myself.

I had a great life growing up and for that privilege, I am incredibly grateful. I'm from a split family (but who isn't these days?) so you know, there was a somewhat complicated family dynamic that unravelled over my childhood. But all in all, I had an incredibly supportive mum and stepdad who both nurtured my passion (shout out to Dad for building me a bird hide down at the dam). I also had a great time with my sisters and younger brother; I can barely remember a time when we'd all be inside.

My younger brother has Prader-Willi Syndrome (PWS), which came with its own set of extra responsibilities as one of his older siblings. In a nutshell, it's a genetic condition that results in developmental, behavioural and sometimes psychological differences. The main characteristic is an intense hunger drive, which can be one of the most difficult aspects for families to manage. Research has indicated that families with a child with PWS experience higher divorce rates and siblings often exhibit moderate to high levels of Post-Traumatic Stress Disorder. Don't get me wrong; my brother is also incredibly funny and infectious to be around. That doesn't negate the pressure that his associative needs place on us as a family. Living in this situation imprinted early that community is a crucial resource for support.

When I reached my teens, realised that I was gay. The confusing realisation came about because I was starting to find people of the same sex attractive, but also because other kids were picking up on it and the bullying had begun. Small/rural towns can be cruel places when you sit outside of the cookie-cutter of what's expected of you. After being called a "f***ot" a few too many times, I made the silent decision that I was not gay (as if that's a choice you can ever make, right?). This marked the beginning of my years denying who I truly was. As an adult, I now realise that that is an incredibly daft thing to do. If you lie to yourself and others about who you really are, that lie is going to completely warp your sense of self and in turn, skew the decisions you begin making. You can't possibly make the right choices if you're not being true to yourself.

The reason that I bring up these aspects of my childhood/teenage years is because they were significant drivers for me finding peace in nature. If you're feeling overwhelmed and you go for a barefooted walk through the bush, how do you feel? The voices/concerns inside your head fade very quickly and are replaced with a chorus of rustling leaves, screaming galahs and white-winged choughs chatting away. I still close my eyes when I hear the bird calls of species from where I grew up and take a moment to listen. It's incredible how these noises stimulate a different part of your brain and help to ground you in moments where you feel anything but grounded.

I moved to Melbourne after graduating high school to study animal & veterinary biosciences. Leaving the country behind was difficult but I felt like I needed to experience new things and meet new people. Although I was living in the big smoke, my views about sexuality had ingrained pretty deeply from my rural upbringing and I maintained the lie of "not being gay". I ended up having a pretty volatile relationship with studies at this point and scraped through with a "P's get degrees" attitude. I had absolutely no idea what I was doing, where I wanted to take my life, who I was or what I was capable of and felt pretty darn lost, to be honest.

Then came the big light bulb moment and I came out at 22. I'd just finished my first degree (which was of no use to me) and my life was about to start taking some major turns. My family was and still is super supportive. I don't know why I ever doubted that, they're amazing people. I think that when you live in an area where there's no LGBTQIA+

representation and you only hear negative comments, you just assume that everyone thinks like that. But even if your family hasn't had much to do with that community, they often become part of it as your ally.

Following this milestone, I realised that I wanted to work in ecology/conservation! Why did it take me this long to realise?! The answer has been there the entire time. But again, you really don't start making the right life choices until you're being completely honest with yourself. This marked the beginning of my journey in the environmental industry.

I thrived in my new degree, a Bachelor of Environmental Science, Wildlife & Conservation Biology. I was hitting high grades and making great new friends who shared my passion. I then travelled to Peru for a conservation project and had begun the long journey of endless volunteering to build my career profile. I ended up extending my studies an extra year to do honours. My project aimed to determine the impact that landscape disturbance (specifically logging and wildfire) has on the yellow-bellied glider (YBG). I was conducting bioacoustic surveys throughout the Central Highlands to determine where the gliders were occurring. If you haven't heard a YBG call yet, I strongly recommend looking it up as it's pretty unique for a mammal! This was a great opportunity to take control of my research project and show myself what I was capable of.

I finished up my project at the end of 2019 and planned to start the job hunt in early 2020. I knew it was going to take a while to land something based on the information I was getting from other early career ecologists/conservationists. Nonetheless, I had hope that the work I had put in (several years of volunteering, a successful honours project, and two years working in vegetation management) was going to be enough to get something at some point. But who could've predicted a global pandemic?!

As I watched the job pool shrink like a dam in a drought, I began extending my search beyond the safety net of capital cities. I've even applied for a couple of positions back in my home region. In doing this, it got me thinking: how do I feel about heading back to a place (or a place like it) where I felt like an outcast? The thought of putting myself in that position does make me apprehensive, but I believe it will be different now. I know who I am, the value I hold as a person, and now have the confidence to stand up for myself. I feel that I won't let potentially negative opinions impact the view that I have of myself.

I think it's important that I remain visible and don't shrink into the crowd like I did when I was younger. That visibility is crucial in signalling to other queer people that I'm there alongside them. There is still an apparent lack of LGBTQIA+ representation in the environmental industry in Australia, alongside a representation of varying races, ethnicities, genders and abilities (I won't speak for the experiences of the latter four because I haven't lived those experiences and that's someone else's voice to be heard).

If you sit somewhere on the spectrum outside of the industry *norm*, trying to find a workplace that champions diversity/inclusivity can further limit a job search. When looking at a job, I can't help but consider if I might experience something negative based on that difference, and if so, whether I will receive the necessary support from my employer if that were to happen. These themes are important for us all to consider. What privilege do you have and how does that impact/advance your employment prospects? I'm still a white straightish appearing male (the straight passing thing sounds weird but it plays a role), so, regardless of me being gay, I still benefit from a hell of a lot of privilege. That's why it's crucial that I act as an ally for others.

Although my job search didn't go as planned, I was still excited for the future! I tried to remain positive about getting a job and not to let the self-doubt seep in. These times are unprecedented and we all need to make sure the expectations that we place on ourselves reflect that. After many applications though, I finally got a job! I've since been working in both consulting and research, both of which have allowed me to go back to rural areas. Being in those spaces for longer periods has felt quite unusual, and I'm regularly reminded that representation of queer people is minimal. That being said, I do believe that some communities need more time to arrive at a place of acceptance and it is an inevitable progression. As more queer people move into/visit those areas, I don't doubt that those communities will begin to understand that our differences are not something to be afraid of but instead are an opportunity for us all to learn from.

# | 35 |

# Science and science communication in the age of social media influencers

## MARÍA ISABEL DABROWSKI

*Let's make the environmental movement focused on science and preservation instead of the perfect Instagram photo, and let's welcome everyone to make a real, true change.*

At a dinner party once I was asked, in front of a group, what I was good at. After a while, I said, *"Well, I'm good at knowing a lot about sea turtles."* People seemed interested and began asking me how many species there are, if had ever seen one and what my favourite *thing* about sea turtles was. And then someone asked, *"Well, are you one of those VSCO-turtle-girls?"* I said, *"A what?!"*

A VSCO-turtle-girl is, for those like me who had no clue, a girl who edits her photos on the popular editing app, VSCO, buys scrunchies and metal straws, maybe dons a *Save the Turtles* shirt and racks up thousands of followers and likes on Instagram.

The question was mainly, I think (read: I hope), asked in jest, but for someone who spends at least an hour on carefully-researched Instagram posts with limited reach, it gave me a lot to think about.

I think this 'VSCO-girl' aspiration is a reflection of a changing environmental movement. It's no longer enough to be passionate, driven, and inspired to conserve and protect for the sake of conserving and protecting. Conservation and the eco-movement are being undermined by people and companies who tout consumerism of the latest (and often most expensive) gadgets, that without, people will feel as if they aren't doing all they can to make a change. Greenwashing, or using hot-button words like 'green', 'eco' or 'natural', trick people into thinking that by buying more, they're helping to save the planet. Similarly, those who haven't yet gone on an Instagram-worthy Caribbean "*eco*"-tourism trip are made to feel as though because they're not in a position to afford a stay at an eco-resort, they can't really afford to make a positive impact on the environment around them. This is simply wrong, and a very damaging side-effect of the age of VSCO-turtle-girls.

In an era where many people are, for better outcomes or worse, jumping on the bandwagon to "save the planet", confessing their love for conservation can make people like me seem like another groupie who finds it fashionable to #savetheturtles. Of course, I have a metal straw – but my posts that feature ocean or bird photos are meant to educate, not to virtue signal or push a product.

I understand this sounds a little spiteful – so let me make two things clear.

First – social media can be a beautiful tool. If you have thousands of followers and have posted about straws, I am not disparaging you – thank you for using your platform to start conversations. I challenge you to keep this up. Beyond straws, talking about ghost nets and by-catch, and how sustainable fisheries can be really hard to find and harder to afford. Talk about the loss of Indigenous and local marine knowledge and practices or about how sea-level rise is slowly sinking islands and flooding communities. Talk about circular economies and second-hand clothing, and how big companies have taken to green-washing to make people continue to buy unnecessary items.

Second – consumerism is a tricky subject. I bought a new shirt the other day, and I kind of love it. It was my first 'unnecessary' purchase in a while (it helps that I am a graduate student with limited means), but I've had great conversations arise when I wear it (it's about coral reefs). I'm challenging everyone to think about 1) reducing consumption and 2) being a mindful and deliberate consumer, for example, supporting small businesses and artists. You don't need 80 different sea turtle bracelets to make a difference. In fact, activities such as volunteering for a beach clean-up, or helping businesses around you find more sustainable practices are much more effective ways to help our planet.

A note I get here often is that really, it's not about the individual. The individual is not responsible for the incomprehensible amount of greenhouse gases released into the air. The individual is often made to feel that way by the true culprits: multinational gas, oil and plastic industries, amongst other behemoth name-brand corporations. For that reason, some people seem content to leave it at that - that it is not the fault of the individual, and that therefore individual action is inconsequential.

I disagree wholeheartedly. Of course, the majority of the fault lies with the corporations who have exploited both people and the planet for decades. But to say then that individuals have zero power goes against the human desire to seek information and, consequently, take action. It comes down to the psychology of change. Get people to care, even in a small way. The decision to use renewable water bottles instead of disposable ones (in places where that is feasible and safe) is a start. People may next start to realize how many plastic water bottles are littered. They may then learn about the breakdown of water bottles into microplastics, and about how the creation of new water plastic bottles can lead to toxic fumes, and how air, light and sound pollution often impact historically marginalized communities. These realisations begin informing not only the lifestyle decisions of the individual, but the decisions of those surrounding the individual. It definitely does not always work like this, but it does, often. These individual voices become a collective, a collective that demands change. So while we must hold big polluters accountable, we should also hold ourselves appropriately accountable (which will look different for each person and community).

I suppose, at the end of this long train of thought, I want this to be the takeaway: individual environmentalism and conservation are not the latest fashion statements. Fashion necessarily goes *out of fashion.* For those of us deeply and determinedly working to actually #savetheturtles, the VSCO-turtle-girl can subvert what we do and who we are, turning the conversation to *"Wow, she's so enviable and gets so many likes"* instead of *"Wow, how can I truly make a difference for our planet?"* Let's not use charismatic mega- and microfauna for a temporary agenda that takes visibility away from the hard work of scientists, researchers and organizers, that takes away from the real problems at hand. We must take action as individuals, but not individuals in a social media microcosm.

# | 36 |

# Coming up stronger

## ANGELA SIMMS

This is a story about my experience with anxiety with my project/ fieldwork in Indonesia.

For those who have done work in Indonesia, conservation is hard. When it comes to the environment, basic conservation concepts are lagging or lacking from the education system (even at a tertiary level), which is quite evident within the community. I have a LOT of admiration for those trying to save species such as orangutans. But this story isn't about my experience with conservation in Indonesia itself: rather about how my mental health declined quite drastically as a result of the unkindness of a person I was working with (just in case you were wondering, they are a "Westerner"; the Indonesians were all quite lovely).

I want to start by saying: I freaking love reptiles. When I was presented with the opportunity to conduct research on a critically endangered, super cool freshwater turtle in a herpetofauna haven, Sulawesi, I was over the moon. My supervisor put me in contact with a person (who will remain as the person/funder through this story), who was looking for someone to conduct a radio telemetry study on the Sulawesi forest turtle.

It sounded perfect. An ecological study that was funded as well. I was very excited and so, got in contact with the person straight away with my interest and to see if this could become a PhD project I could pursue. They were based overseas, so contact was through email, Skype and Facebook messenger. I set myself up at a university with two more amazing supervisors (although it was in another state, so contact was pretty limited for some time). And we were set and then waiting for scholarship rounds to open at the end of the year.

I maintained regular contact with the funder throughout the six months (maybe more) to help on the project where I could before starting my course. I would help write up or proofread grants and with setting up permits for myself. For the most part, it was fine on my end, but my partner could see things turning for the worse as my relationship with this person progressed, although I was far too insistent and stubborn that everything was OK. I was set to go on my first trip to Sulawesi before the university semester commenced.

October 2018. The night before I flew out to Palu, an earthquake hit the city, triggering a tsunami and turning the ground into mush (a phenomenon known as liquefaction). I was in shock. I never prepared for a natural disaster (I live in Melbourne, it's pretty safe here). I was actually meant to leave a week beforehand, but because my permits were delayed, we delayed my flights by a week. I would have been at a hotel that got destroyed by the tsunami if it wasn't for that permit delay.

This is when I started to notice my anxiety. Despite this, the funder still insisted I fly to Jakarta for the week to sort out permits. There was also an immense amount of pressure from the funder to still fly to Sulawesi despite what had just happened; thousands of people had lost their lives, and many more were displaced. I later saw the sheer destruction it had on the city and while hearing the stories, I was always holding back tears. It was devastating. Thankfully, all Indonesian academics and my supervisors said a firm no to me going anytime soon.

Over the next few months before the follow-up trip, I began having bad anxiety and what I now recognise as panic attacks. The funder's personality in these months was stronger than ever. I was warned about the person's personality from sources on my trip to Jakarta, and I saw the stress they projected; I forwarded the source's messages to my supervisors, but we were all still positive the project would be fine and that the funder would calm down.

Finally, the trip the funder had been pushing for so badly came round, although on quite short notice to get the paperwork sorted through the university (and honestly, it was still early days from the disaster). What should have been a red flag was me crying the night before because my partner and the funder had different ideas (funder wanted me to bring a tent over to sleep in somebody's yard, despite having said there is accommodation. My partner and even my dad said *"Hell no."* I'm glad they stepped in for my safety – on my second trip, I found out a neighbour/friend had been murdered with a machete to the head one night – and, as my research assistant put it, this is a scary village). I was still optimistic that the funder was better in real life than over messenger.

But it got worse. I felt more anxious than ever being around this person. It was so bizarre: never had this happened to me. There was a lot of things that I didn't agree on with the way this person was running the project, such as the lack of respect and professionalism this sort of project needs with local counterparts and actual experts (not just the expert they saw in themselves); however, there was a lack of openness for discussions to a point where I just stayed silent to avoid a volatile and over-emotional response from the person.

After five days, they left to go back to their country. On that final night, I was in bed, trying not to cry as we shared a room (I had tried to stand up for myself/supervisors after a comment they had made) – I then got woken up abruptly at 4 am before their flight, to be told to *"make sure you take photos"*, not to say goodbye or good luck for the next two months. I find that part funny now, at least. Later that morning, it was the first time I cried in front of a supervisor over the phone (completely embarrassed thinking how unprofessional it was to do so, and thoughts running through my head – *"am I being crazy/irrational?"*- I did not intend to cry at all. I was just calling to check-in), he handled the situation with such grace and we were still optimistic about the project.

Two of my supervisors came over to Sulawesi to help set the project up a few days later and my goodness, did they help my mental state! It wasn't until they saw the Facebook messages from the funder on the trip and turned to me and said *"Ang, this is harassment"* that I finally recognised where the anxiety and panic attacks stemmed from; not feeling I was doing good enough, based on this one person's *very strong* opinion.

From then on, my supervisors stepped in and began to shield me from this person. They'd even avoid using her name around me (to make myself laugh, it reminded me of *Harry Potters'* He Who Must Not Be Named – Voldemort). I recognise now and would describe this person as over-reactive, overemotional in a professional sense, with a lack of respect towards people and quite frankly a self-proclaimed biologist/expert who was rather ego-driven (with no formal study or training); an arrogant and strong personality, which made it way too easy to belittle me and tear down my self-confidence. Despite my supervisors stepping in, unfortunately, the next two months weren't easy as there were a few more challenges.

I had a lot of battles in my head due to the funder, but my experience on this trip gets worse with a hospital experience – another experience that shook me up was watching a 45-year-old man (who was my guide) go into a complete meltdown at 6:30 am when his wife passed out from the pain of appendicitis – he thought she had died. Trying to get her into a recovery position and check her pulse alone was impossible as he continued to violently shake her – a first aid course does not teach you how to calm down a hysterical person. I couldn't calm him down. Nobody had any phone reception for me to call the university's emergency line and we were 4 hours away from the nearest hospital (and had no car). It was a nightmare – and the actions that followed for the next four days - which I put down to lack of education - were easily some of the most frustrating moments of my life. The lack of common sense was mind-blowing and literally life-threatening. But that experience was easier to process and get over than the harassment I had experienced from the funder.

I have only skimmed the surface, but I completed the fieldwork. The best way I could describe the fieldwork was – every day I felt like I was trapped in a nightmare with what was running through my head, where sleep was an escape. I had a lot of dreams about being back home. I came back feeling quite broken, my confidence shaken, feeling immensely guilty as my supervisors had to step in like that (and having used their own funds to come visit me in Sulawesi, I didn't want to disappoint them), I kept thinking of what I could've done to have avoided that whole experience with the funder, wishing I could turn back time.

With the help of a psychologist, my incredible partner, friends and family and three absolute superstars of supervisors, I am coming to terms with those feelings. I get so nervous so easily and have cried at the most simple questions – like my supervisors asking me, *"How are you doing?"*; I'm even crying now just writing this story. I've had moments where I could feel my heart thumping against my chest so hard and abnormally, and I've had a really hard time identifying triggers before I panic or cry (also to note: my "normal", is a LOT more chilled out and chirpier than this, this project has completely transformed a part of me). But that's okay, I am working through it.

I am now on my second surveying trip for six weeks and although the feelings from the past are creeping in with memories being resurrected, it's good, I'm facing an experience that has hurt and scared the hell out of me and turning it into something better.

I am grateful to have learnt a great deal from this experience and what I can do in the future to prevent it from tearing me down again. As one of my supervisors put it, my soft/sensitive personality is a quality he likes about me, and I don't need to change that. It has also taught me how vulnerable we are in these learning positions, and I was incredibly lucky to have such progressive, nurturing supervisors (I know that isn't always the case).

Bullying and harassment have no place in conservation and science, and if you ever face the minority that belittles you, makes you feel inadequate or anxious in this field (it can sometimes be incredibly hard to recognise as I've found out) - I'm not saying you need to necessarily stand up to the bully (I've seen the funder be called out for their behaviour, they're not going to change and I accept that some people won't) - but instead, surround yourself with positive reinforcement. There is so much extra support out there (including this *Lonely Conservationists* group!). In my personal experience, it was quite easy to be open with my supervisors about my mental health and what I felt most anxious about. Speaking to my partner and friends about upsetting events also helped me get through a lot of the emotional feelings that I had. I have a mental health plan from my doctor, which has enabled me to get free psychologist appointments as well (as for many conservationists, money is often an issue. Also finding a good psych for myself, that's a whole other story, but do not give up!). Do what you need to do to bring your confidence back up to kick some goals in conservation. Personally, I think you sometimes need to be knocked down to come back a hell of a lot stronger.

Stay kind, conservationists.

### Final Note:

I have decided to not continue this project specifically as a PhD, it will stay as a Masters project. But this has not deterred me from a life of science, research and ecology whatsoever! I am currently exploring other projects to take on, and I am incredibly excited for the future!

# | 37 |

# Never Lose Hope

## NENE HAGGAR

Indonesia isn't all beauty. It isn't the poetry of wildlife you hear echo throughout the forests but rather, the sound of chainsaws annihilating them. It isn't the smell of fresh air in the morning that surrounds your existence; it's a smoky cloud hovering above you as flames dance and burn landscapes to the ground. These cleared landscapes now stand tall with plantations of oil palm. Spreading alarmingly, unsustainable palm oil has taken all of Indonesia's wildlife along with it.

Throughout the rest of the world, there is a calling- a demand to cuddle an orangutan; to take a selfie with a stupefied tiger; to eat the flesh of a pangolin and rob them of their scales. Indonesia responds to this calling: babies are taken away from their mothers and exploited into the illegal wildlife trade each passing day.

Some, such as myself and my colleagues, travel to Indonesia with the hope that they can contribute to restoring its pristine forests. However, placing yourself in Indonesia, you witness all of its destruction. For many of us, being a bystander to the deteriorating state of the environment can be overwhelming; you can feel consumed by it and left feeling numb. We may be left with no direction, not knowing how to take action because we've fallen into despair.

While being here, I've crossed paths with people who have lost hope. Many have accepted that their children will never see an orangutan in the wild. Are they right to say so? Yes. If no action is urgently taken then orangutans, along with Sumatran tigers and other charismatic species, will become extinct.

However, despite the blood, sweat and tears of being in Indonesia, I still believe that we can bring these species back from the brink of extinction. I still have hope.

I have hope because endangered species have bounced back before, although this doesn't happen on its own accord. It happens because of the hard-working people who dedicate their lives to saving such species. They devote much time and effort to the cause. Time is something humans often find difficult to grasp but with just a little patience, people can create change. When you look back at the people who passionately fought to save the black-footed ferret, it took decades to do so.

Success stories are usually generations in the making. If we didn't have conservationists who kept pushing forward, then we would probably live in a world without rhinos or California condors. Many animals might only exist in zoos; imagine the extent of species that would have already vanished from this planet forever! The people who fought so hard to preserve these species believed in what they were fighting for, they never gave up. These people always had hope and that, in itself, is beautiful.

So, my message is that no matter how doubtful or pessimistic you may feel at times, never give up hope because, without people like you in this world, the planet would be a very different place. You and your drive are what makes conservation work so well; it's literally how we save species from extinction! As Dr Jane Goodall once said:

*"You cannot get through a single day without having an impact on the world around you. What you do makes a difference and you have to decide what kind of difference you want to make".*

Today, there is a very small portion of forest left in Indonesia. But that forest, and all of its beauty, is worth saving. I believe that if we all work together and support one another; the animals of Indonesia (and the rest of the world) will win the race against extinction.

# | 38 |

# Saying yes works great (until it doesn't)

## GILLIAN

When I was 10 years old, I decided one day that I would no longer be afraid of anything. Spiders no longer meant me harm, heights were just temporary discomfort, and the shadows in the forest should be explored- not feared. All I had to do was just push through my hesitation *and just say yes to everything*- and it all would be fine. Simple, right? I thought so at the time. My new mentality pushed me away from the painful shyness that had defined my childhood and towards exploring the wilderness in my hometown, pursuing sports, and standing up for myself.

As a 13-year-old, I solemnly concluded that it was time to prepare for college and a career. Having grown up surrounded by pets and wildlife, pursuing wildlife biology seemed like the obvious choice. I wanted to start my career as soon as possible, but the tricky thing about being 13 years old is there's a lot of growing up to do before anyone lets you, you know... perform research on dangerous animals and whatnot. To bypass this barrier, I began volunteering at the local zoo, saying yes to every opportunity along the way.

*"The zoo is pretty busy today, can you handle this large crowd?"* Yes.

*"The penguins bite, are you sure you want to go in there?"* Yes.

*"We could use more help on this holiday, can you do an extra shift?"* Yes.

Saying yes often brought me fun experiences and knowledge, even if sometimes it took up time that I didn't have. Yet I cherished my time there, and it wasn't long before I found my niche- big cats! I spent all the time I could working near them. During my shifts, I soon noticed a pattern: the zoo's visitors could recognize and recite facts about lions, tigers, and leopards, but often knew nothing about mountain lions. Some didn't even know that they were here in California. That's when it hit me: I wanted to work with big cats, and there were some practically in my own backyard that needed to be understood better.

At the age of 19, I took my next step and began interning with my college's mountain lion research project. I always said 'yes' to carrying the heavy supply packs on long hikes, traversing perilous canyons and valleys, entering burn zones, and working during heat waves and storms alike- even when I felt uncomfortable. I told myself it was what biologists did, and if I didn't do it, someone else would. I learned how to navigate, place trail cameras, use telemetry, and track. I assisted in more than two dozen capture attempts and helped to sedate and collar several mountain lions. I'll always regard these years of research as some of the most fun and formative of my life.

Yet despite finally feeling like I was doing what I was *meant* to do, doubt towards my abilities and a fierce need to be taken seriously always lingered in the back of my mind, unspoken but constant. The need to prove myself in both my college life and my internship usually left me working myself to exhaustion. I felt like saying no to opportunities was evidence that I wasn't a "real" biologist, college student, or athlete. At one point I was working two jobs, two side gigs, the internship, five-six days a week of NCAA track and field practice, and a full college course load.

More than once I studied for midterms in a tent using a flashlight during overnight captures, and would be exhilarated but exhausted the next day (*Worth it? Yes. Did it impact my scores? Likely*). When I would return home from some extra-long field days covered in mud and blood, I secretly relished the horrified looks I got from housemates and would often joke about the times I almost got heat stroke during long hikes or broken bones sliding down some steep terrain. Online, I saw other biologists proudly sharing their similar field horror stories and the unbelievable stress they were under in school or work, and felt validated. People praised me for working so hard and putting myself out there, feeding the beast within that associated being reckless with working hard.

One morning, when I was supposed to go out by myself to pick up a trail camera, some rain rolled in. My supervisor emailed saying that he would appreciate it if I still went out, but understood if I didn't want to. He is a kind and fair guy, so in hindsight I know he wouldn't have minded if I'd said no. But at the moment, reading that email made me feel like it was another challenge I shouldn't say no to. My imposter syndrome and the need to prove myself as a tough biologist overrode the common-sense feeling in my gut screaming that it was a horrible and dangerous idea.

I set out into the rain. It was freezing, but I had brought multiple sets of clothes, warm coffee, and rain gear. I had hiked in the rain many times before and naively thought this would be the same. It was not. I knew that once a biting wind picked up and I began to lose feeling in my fingers. I kept going. My logic was that I had hiked so far that I might as well accomplish what I said I would do, secretly afraid to say I failed the task if I turned back.

After I reached the camera, I was soaked to the bone, and my teeth began to chatter uncontrollably. That was when my brain started getting increasingly foggy. My legs grew heavy as I stumbled back down the trail and I lost feeling in my tongue. I began to mutter nonsensically to myself as I just kept trying (and failing) to put one foot in front of the other. At one point, I sat cross-legged in the mud for a few minutes, trying to summon some energy to get back to the parking lot. For some reason, the most frightening part of it all was the strange, deep longing I felt to crawl off the trail and curl up in a hollowed-out redwood tree to sleep. It was all I could think about.

Somehow, despite a glitchy GPS, no phone service, and my brain shutting down from what I assume was some level of hypothermia, I finished the long trek back to the parking lot and then drove an hour home in the dark. How I managed that, I have no idea- I was not in any condition to drive, but at that point, my brain had turned off. I was shaking violently (the truck's heating was broken) and only remember flashes of it to this day.

After taking an hour-long warm shower and sleeping for most of the next three days, I finally had a clear head again and realized just how much I messed up. My insecurity had put me in danger and I vowed it would never happen again. It took a lot of time and self-reflection to work past the anxiety I would feel when hiking alone after that day.

The experience was a huge wake-up call. Since then, I have worked on never saying yes when my gut or my mental health tells me no- and I am a better biologist for it.

I've learned that working hard will get you far (after all, catching a mountain lion isn't usually easy), but adding self-confidence to that gets you farther. I'm proud to say that I'm now a meticulous hike planner and trail safety extraordinaire, and I actively work on trusting my abilities and instincts. To anyone reading this, I hope you find it in yourself to define the line between working hard and being self-destructive. Imposter syndrome runs deep, but I believe in you. Hopefully, with practice, you will too.

# | 39 |

# Listening to Pain

## ANNABEL

### The Beginning

My childhood was fairly typical of a privileged conservationist: I grew up surrounded by nature and had the opportunity to pursue my passion with a supportive family and access to a good education. But when I turned 17, my life changed.

I developed chronic pain.

My pain started, for some unknown reason, just before my final year of high school, universally considered the most important and most stressful year. The strain of that year on my nerves and muscles caused constant arm pain while doing simple tasks like writing or holding a book open. On some days, I couldn't even brush my teeth or button up my shirt without searing pain racing through my arms. Yet every doctor said it would naturally heal because I was so young.

One of the hardest parts was feeling that I had lost myself. Everything I once loved – studying, drawing, reading, making music – I could no longer do without pain. If I could not work with my arms, then did I have any value? How could I even pursue my conservation dreams?

Although I had prepared myself for a terrible university entrance mark, I ended up receiving the exact mark I needed to gain entrance into my first choice. Over the moon, I thought I could finally do what all those doctors had been advising me to do – rest! But instead of healing my pain, the long period of rest had just deconditioned my muscles. I started university with muscles that could not weather the strain of study and a nervous system that was wired to fire pain signals.

## The University Years

My first year of university was pretty brutal. I spent the first four weeks of the semester in pain, the middle five finally coping and the final four again in agony. By exam time, I could not type or write out study notes as I had to save my strength for the actual exams. Instead, I would just read over my lecture notes and create mnemonics to help remember the content. In second year zoology, I even created a mind-palace to help memorise key facts about all 36 animal phyla!

Despite the obvious disruption the pain was causing in my life, I struggled to find a sympathetic doctor. During my first two-and-a-half years of chronic pain, I saw a total of 10 different medical and allied health professionals. I would end up crying to each new health professional in fear of being dismissed again. Finally, though, I found a sympathetic GP who sent me to the local pain clinic, a specialist upper-limb physiotherapist who put me on track to strengthen my arms, and an amazing osteopath who relieved muscular tension. I also found a psychologist at the university counselling centre that had worked in a chronic pain clinic. He introduced me to a fantastic book by Norman Doidge *The Brain that Changes Itself,* which has a chapter on pain psychology (and was later expanded upon in his next book *The Brain's Way of Healing).* And finally, I had my parents, a constant source of support – I honestly could not be where I am today without them.

## The Conservationist in Training

For my honours research year, I had two main goals; to learn more experimental design and develop my fieldwork skills. And I'm so proud of what I achieved that year: I completed three field experiments, learned to kayak and drive a boat, learned basic R coding, wrote a 60-page thesis (not including figures and references) and endured a sting from the famous Australian giant stinging tree! Thankfully, I was also able to work for my honours supervisor as a research assistant for 10 months.

The next difficult step was finding a job, a subject that has been thoroughly discussed in *Lonely Conservationists*! I would like to add two perspectives of mine to this topic.

One of the difficulties in having chronic pain is that it rules out some early-career conservation jobs. I have to be careful to not cause a flare-up with strenuous physical labour, ruling out a common stepping-stone job for conservationists in Australia (field officer). Repeated movements can also cause a flare-up and it's stressful to tell colleagues when I need to rest.

Last year, during my conservation internship, I had the opportunity to assist in potoroo trapping in a beautiful part of southern Australia. About halfway through the set-up day, my pain was flaring up from carrying the traps. Thankfully, my friend and fellow intern noticed I was struggling and knew that I wouldn't ask for help for fear of letting the team down. It meant the world to me when she quietly took some of my traps and carried a heavier load.

It's also really scary telling a new employer about my chronic pain. One of the first interviews I had after finishing university was for a graduate position at an ecological consultancy. I was very excited until I read the pre-interview questionnaire. In it, the company asked if I had any medical conditions that might affect my work. They also stipulated that if a successful applicant was dishonest in the questionnaire, they could terminate employment. In fear of retribution, I wrote down my chronic pain and it was, inevitably, brought up in the interview. Afterwards, I learnt that in Australia, there is no legal requirement for me to tell a potential employer about my condition. And if I don't inform them, the worst outcome is that I might not be able to take workers' compensation if my condition worsens due to the job.

Although I know I'm not defined by what I can do, sometimes, I feel like I'm just damaged goods. That interview definitely made me feel like that.

## Chronic Pain

I've never wanted to be defined by my chronic pain. Even though it affects my ability in life, it doesn't seem like a real disability. Chronic pain is such a nebulous concept, an ill-understood condition. If it was more definable, maybe I would be more deserving of assistance. One in five people experience a chronic pain condition and yet it was only added to the medical diagnostic Bible, the International Classification of Diseases (ICD), in 2019 for the ICD-11. My pain is here to stay and I have learned to pace myself; to manage it rather than fighting it.

Chronic pain taught me to listen to my body, conservation taught me to listen to the earth.

# | 40 |

# On Imperfection

## JON KAHLER

I can't say that I had a very unique upbringing.

Raised in the sunny suburbia of Brisbane City, there was nothing exceptional about my childhood. My family was never very environmental, and between my ten fingers, not one of them was green. While it might make people who know me wonder how I ever became so invested in environmentalism, my parents always placed a strong emphasis on right and wrong, good and bad. Like any child, this was instilled in me from a young age and this strong moral compass has directed me forward throughout life.

I remember learning of the effects of global warming as a young child and writing a letter to the Australian Prime Minister at the time, John Howard. Seven-year-old me had an amazing plan to create a machine that mimicked trees and sucked all the greenhouse gas out of the air, instantly solving the climate crisis. You can't say I wasn't ambitious. A few weeks later, I got a stock standard reply talking about the *aspiration* and *innovation* of future generations. My dream to single-handedly fix climate change was halted.

Flash forward to my late teen years, and I found myself reading more moral philosophy. People were putting forth frameworks of how to do the right thing and live good lives. These weren't mere intuitions, but strong argumentative stances of what it means to be a good person. It was here that I was exposed to the power of ideas. Not just theoretical ideas either, but the tangible change that is possible through applied ethics. From reading these works and seeing those who were willing to speak out against injustice, I was inspired by the possibility of true ethical change within our society. Not a convenient change either, but fundamental shifts that questioned the norms of our society. I knew that through ethics people could be motivated to change their habits and be inspired to care for the natural world. This is what motivated me to pursue environmental ethics.

The word *ethical* is used a lot nowadays. We hear of ethical brands, ethical fashion, ethical influencers and ethical businesses – a mostly contradictory term, in my opinion. I think many people use it, in part because they want to believe it is true. No one wants to be unethical. I know that is true for me at least. I want to strive to be an ethical person even though I know I will fall short. However, it took me some time to come to this realisation. Heading down a career of ecological ethics, I came to understand that acting ethically is not easily done; being an ethical person is even harder. For a long time, I was hoping that studying ethics would enable me to be a more moral person. However, I quickly came to realise that studying ethics is no guarantee of moral character. When looking at the demanding nature of any appealing ethical theory, it seemed hard to imagine myself upholding it. Sure, some people buy clothes made of hemp, don't use single-use plastics and have given up international travel. But to be truly ethical and devote oneself completely requires a sacrifice most aren't willing to make.

For some time, this plagued me. Despite studying a degree that emphasised both the individual and societal change necessary to create a more sustainable world, I was not able to act on those ethical conclusions. Here I was asking 'how we *ought* to act', and then not following through. Instead, I found myself criticising my every action.

Any time I did something remotely anti-environmental, I would feel dejected. Every time I used a single-use plastic, bought something unnecessary, didn't ride my bike or wasn't campaigning, I had this sense of shame that I could be doing *better*; I could always do *more*. I think this is something that any environmentalist can relate to. For each waking moment, you're working against a clock. As time ticks towards doomsday, each minute is a loss. Habitats gone, species pushed further to extinction and pollution continuing to plume. It is easy for anyone in the environmental sector to see time away from work as time wasted. Or dare I say, unethical. Naturally, this can lead to stressful, self-destructive, or even counterproductive habits. And so, we must first look after ourselves, before we begin to tackle the larger issues of environmentalism.

For me, that meant coming to terms with ethical imperfection. I was never under the impression that I was morally infallible. But understanding how hard it was to even strive for an ethical life was tantamount to pushing a boulder up a hill.

However, there is something to be said about the imperfection of our ethical decisions. It is only through those imperfections that we can more readily decide to be better. If we don't question our actions – and more importantly learn from them – how can we ever expect moral change to occur? It is through our mistakes and struggles that we learn to be more moral people. Whether we be ecological ethicists, conservationists, marine biologists or environmental activists, we can be humbled by our imperfection. Submitting ourselves to this fact can allow us to start acting with greater insight, not only in our moral life but also in our professional life too.

Since realising my limits – both physically and mentally – I have been able to make the best decisions in the now, for greater return in the future. When I feel defeated or I feel the need to remove myself from the insurmountable odds that environmental work poses, I can be content with that choice. I now know that in saving myself today, I can do my best to save the world tomorrow.

# | 41 |

# For the sea

## ANA WILLIAMSON (NÉE WILLETT)

I think out of anyone, I understand impostor syndrome on such a deeply profound level that it's disturbing.

My name is Ana, and I am a marine conservation policy contractor. There...for lack of better phrasing, is not a lot of work in this field in the D.C. area. You either have to have a PhD (I don't, I have a Masters), or you have to know someone. It leaves a lot to be desired when all you want to do is make an impact and save our seas. It makes it difficult, with such a desire to protect, to feel stuck and unable to do much at all when there are opportunities that seem to skip by you.

I have always loved the sea—desperately. Immensely. It has always felt like it has called to me, whispering a siren's song of *"help me"* and *"protect me"* from its coastlines when I would go to the beach as a child. I would, as silly as it sounds, go sit at the edge of the shore, letting the waves lap at my feet, and idly chat with the ocean as if she were an old friend of mine. When I was five years old, a fisherman had caught a baby shark off the Rodanthe Pier in the Outer Banks of North Carolina. I was so horrified that he was not going to release it back, and treated that poor shark as a trophy, that I begged my mother to ask him to release it to the sea. He did, and I'm fairly certain it was because there was a five-year-old who wouldn't stop crying.

I grew older and my love of the sea remained. Eventually, I went to college, studying international affairs and World Languages with a focus on conservation policy and Romance languages. My university, up in the mountains of West Virginia, definitely did not have a course schedule that allowed me to truly pursue my love of marine conservation, but diligent as my professors were, they found a way to make it work for me.

When I had graduated from undergrad in 2016, we had just had the upheaval of a lifetime. Donald Trump had been elected President of the United States—the EPA was ripped to shreds along with my post-grad opportunity with them, anything remotely akin to conservation policy jobs was swept away as quickly as the tide goes in and out.

So I went back to school, and even after two years, I was still struggling to find a job in my field or even somewhat close to my field. It wasn't until 2019 that I was hired by a small nonprofit organization in Washington, D.C. to do something nowhere near what my background is. But I needed a job, desperately. So I took it. And I'm still there.

But what was earning me money left me deeply unsatisfied, unhappy, and feeling as if I had failed; I wasn't doing the work I was born to do, I wasn't writing policy briefs on marine protected areas or on how osprey migration patterns have shifted in the past five years—due in part to a change in Chesapeake Bay conservation policies in Maryland.

I felt, and still do feel, stuck. So I took up contracting as a moonlighting side gig. There, too, is not a lot of work in contracting in Washington, D.C. Occasionally, I'll get a request to review policies on marine protected areas, usually for large corporations. But again, it leaves me feeling desperately unsatisfied and unhappy with what I'm doing.

The point of this is: I know I'm a specialist in marine conservation policy. I know my talents will be needed somewhere, somehow, someway. And whether I find that opportunity today or in five years, I know I will still love the sea, still have a deep, immense desire to protect it, and continue to fight for it, even if it's not in the way that I imagined.

If, for the time being, my fight for our coastlines and waterways is limited to me doing monthly beach clean-ups, oyster planting in the Chesapeake Bay, and volunteering with local nonprofits that advocate for our seas and waterways, then so be it. I know, as long as I keep chipping away, that I will make it.

Despite all these hiccups, despite the ups and downs, the hurt and frustration, what we do is remarkable. We're warriors and advocates for our Earth—and we only have one. If we don't fight for it, who will?

# | 42 |

# The forever field assistant

*NATASHA BARTOLOTTA*

With my shirt sticking to my back and muddy water spilling into my boots, I trudged through the swamp forest of Sumatra deciding, once again, to extend my time at an orangutan research field station. I had been there for 15 hot, humid months and had already extended my stay twice.

First, I didn't want to leave because I loved it so much and was asked to stay. Then, I offered to stay longer again because the project I was working on could use more data. Now, I was asked to remain just a tad longer to help train new volunteers. I always thought, *"Why not?"* This thinking is what turned my post-undergraduate gap year into three. Two of those years had been spent as a volunteer research assistant collecting data on great apes.

At some unknown point during my childhood, chimpanzees became my favourite animals. I specifically remember watching the 1997 movie *George of the Jungle* and hearing a giant, talking ape *(naturally)* say:

> *"Madam, I knew Jane Goodall, and you are no Jane Goodall."*

Mini-me wondered, *"Who's Jane Goodall?"* Learning her story blew my little child mind. I was amazed that a woman had actually gone on her own to study these animals. I didn't know a person could do this as a career, but now, the seed had been planted in my brain and begun to take root.

Following my unique path full of wild field biology experiences, I eventually achieved my dream of seeing chimpanzees. I had just graduated college with my biology degree and I was off to Uganda for my first extended volunteer research position. My glorious purpose here was to catch chimp urine and take faecal temperatures. As I got inadvertently peed on by a chimp for the 50th time (and pooped on, too, many other times), I wondered if this is what Jane felt like all those years ago.

Unfortunately, my time there ended a few months early because of a personal mistake I made. Something that (without going into the details) left me deeply embarrassed and discouraged. I saw this as a very personal failure and that I had ruined a rare opportunity. However, the most important part of every life lesson is to move forward and better ourselves with each step we take. So, move forward I did.

As fate turns out, leaving the chimp site early allowed me to travel to Indonesia. I really thought I would apply to grad schools when I came home from Uganda, but then I saw a posting for another volunteer field assistant position and again thought, *"Why not?"* Little did I know, I would end up falling in love with everything about this place: the orangutans, the community of people I met, the research, and even the itchy, swampy forest. My life goal so far had been to become Jane Goodall. I thought I wanted to study chimpanzees, not orangutans. Now what? Maybe I will just have to do for orangutans what Jane Goodall did for chimpanzees!

This wasn't the only identity crisis I faced. Coming back home, I felt like I had spent the last three years of my life stuck in a perpetual *in-between* phase. I'd seen my friends and undergrad peers move forward with their lives, get paying jobs, get their own places, get into grad schools, and start working on their own research projects. I too, am hoping for the day I can call a research project my own. I met many master's students my age in Indonesia and I couldn't help but feel a little behind them as I continued to collect data for someone else.

But you know what? Everyone moves at their own pace and walks their own path.

I am planning to go to graduate school in the fall for a Master's within the orangutan research project I was just volunteering for. So, I may finally be moving forward in my education, but I'm still looking forward to the day I have a research project that is entirely my own creation from start to finish. When I think about what I want to research, I am also conflicted. Should I choose a topic that will reveal something about great apes just for the sake of knowing it? Or should I strive for my research to have specific conservation implications? Nonetheless, the better we understand a species, including all aspects of its ecology, behaviour, and life history, the better we can implement successful conservation strategies. Thus, all researchers are conservationists.

Even more, scientists who share their research and their passion for their study species can raise awareness and spark empathy for endangered wildlife. This is why I am now getting into the science communication world through creating my blog, sharing my research experiences on Instagram, and speaking to high school students. I even second-guessed myself when I had to put a label on my Instagram account. Can I call myself a scientist? Officially?

Yes. I've done everything a field biologist does, just as a volunteer. I am a field biologist. I am a researcher. And I am a conservationist. Most importantly, I am also myself. Perhaps, I should no longer focus so much on becoming the *"orangutan Jane"* as I should on being whole-heartedly Naturalist Natasha.

# | 43 |

# Acknowledgements

This book, in its entirety, was produced by the *Lonely Conservationists* community and I would like to thank each author for being an absolute pleasure to work with. As well as writing their pieces, the authors worked through the editing process independently and were involved in every stage of producing this book.

I would like to thank Renuka Kulkarni for conducting the final manuscript edits while still maintaining the individual voices, language and speech mannerisms of the authors. It was important to this project that the individualism of the author's stories shined through from start to finish and I was so fortunate to have Renuka helping me to balance the author's verbal flair alongside proper grammar and readability.

I would also like to thank Jack O'Connor for the cover art and Phalguni Ranjan for the title illustrations. There are so many artistically talented individuals within the community and I am proud to have their creativity represented amongst a sea of scientifically talented individuals. It goes to show, you don't need to pick a side and that you can be scientifically and artistically minded in unison.

Lastly, many thanks need to be allocated to the community as a whole for supporting and uplifting each author who is brave enough to publish their story. The support of the community has changed the lives of many conservationists around the world and without it, this book would not exist.

***

If you too feel like a lonely conservationist, you are more than welcome to join the community over at www.lonelyconservationists.com for weekly stories, social media updates, mental health resources, the podcast and more.

Funding a global community that intersects industries and spans nations is a challenge, so if you, your business or organisation would like to support *Lonely Conservationists* and our projects, please contact jessie@lonelyconservationists.com because we would love to work with you!

www.ingramcontent.com/pod-product-compliance
Lightning Source LLC
Chambersburg PA
CBHW072123020426
42334CB00018B/1691